Great Science Adventures

Discovering the Ocean

Great Science Adventures is a comprehensive project which is projected to include the titles below. Please check our website, www.greatscienceadventures.com, for updates and product availability.

Great Life Science Studies:
The World of Plants
The World of Insects and Arachnids
Discovering the Human Body and Senses
The World of Vertebrates
Discovering Biomes - Earth's Ecosystems

Great Physical Science Studies:
The World of Tools and Technology
The World of the Light and Sound
Discovering Atoms, Molecules, and Matter
Discovering Energy, Forces, and Motion
Discovering Magnets and Electricity

Great Earth Science Studies:
The World of Space
Discovering Earth's Landforms and Surface Features
Discovering Earth's Atmosphere and Weather
Discovering Rocks and Minerals
Discovering the Ocean

Copyright © 2006 by:
Common Sense Press
8786 Highway 21
Melrose, FL 32666
(352) 475–5757
www.greatscienceadventures.com

All rights reserved. No part of this book may be reproduced in any form without written permission from Common Sense Press.

Printed in the United States of America
ISBN 1-929683-23-5

The authors and the publisher have made every reasonable effort to ensure that the experiments and activities in this book are safe when performed according to the book's instructions. We assume no responsibility for any damage sustained or caused while performing the activities or experiments in *Great Science Adventures*. We further recommend that students undertake these activities and experiments under the supervision of a teacher, parent, and/or guardian.

Great Science Adventures

Table of Contents

1. What do we know about Earth? 6
2. What do we know about water? 8
3. What do we know about the ocean? 12
4. What is oceanography? 14
5. What is saltwater? 16
6. What do we know about the ocean floor? 20
7. What are the ocean layers? 22
8. What lives in the ocean layers? 26
9. What are waves? 28
10. What are tides? 32
11. What are ocean currents? 34
12. How does the ocean affect weather? 36
13. What is the intertidal zone and what lives there? 38
14. What do we know about life on the seashore? 40
15. What are estuaries? 42
16. What do we know about crustaceans, mollusks, and sponges? 44
17. What are the three types of fish? 46
18. What do we know about anemones, marine worms, echinoderms, and marine fishes? 48
19. What are coral reefs? 50
20. What do we know about sharks and rays? 52
21. What do we know about marine reptiles? 54
22. What do we know about marine mammals? 56
23. What do we know about pinnipeds? 58
24. What are ocean resources? 60

Lots of Science Library Books 63

Graphic Pages 143

Great Science Adventures

Introduction

Great Science Adventures is a unique, highly effective program that is easy to use for teachers as well as students. This book contains 24 lessons. The concepts to be taught are clearly listed at the top of each lesson. Activities, questions, clear directions, and pictures are included to help facilitate learning. Each lesson will take one to three days to complete.

This program utilizes highly effective methods of learning. Students not only gain knowledge of basic science concepts, but also learn how to apply them.

Specially designed *3D Graphic Organizers* are included for use with the lessons. These organizers review the science concepts while adding to your students' understanding and retention of the subject matter.

This *Great Science Adventures* book is divided into four parts:

1) Following this *Introduction* you will find the *How to Use This Program* section. It contains all the information you need to make the program successful. The *How to Use This Program* section also contains instructions for Dinah Zike's *3D Graphic Organizers*. Please take time to learn the terms and instructions for these learning manipulatives.

2) In the *Teacher's Section,* the numbered lessons include a list of the science concepts to be taught, simple to complex vocabulary words, and activities that reinforce the science concepts. Each activity includes a list of materials needed, directions, pictures, questions, written assignments, and other helpful information for the teacher.

   The *Teacher's Section* also includes enrichment activities, entitled *Experiences, Investigations, and Research.* Alternative assessment suggestions are found at the end of the *Teacher's Section.*

3) The *Lots of Science Library Book* are next. These books are numbered to correlate with the lessons. Each *Lots of Science Library Book* will cover all the concepts included in its corresponding lesson. You may read the *LSLB* books to your students, ask them to read the books on their own, or make the books available as research materials. Covers for the books are found at the beginning of the *LSLB* section. (Common Sense Press grants permission for you to photocopy the *Lots of Science Library Books* pages and covers for your students.)

4) *Graphics Pages,* also listed by lesson numbers, provide pictures and graphics that can be used with the activities. They can be duplicated and used on student-made manipulatives, or students may draw their own illustrations. The *Investigative Loop* at the front of this section may be photocopied, as well. (Common Sense Press grants permission for you to photocopy the *Graphics Pages* for your students.)

Great Science Adventures

How to Use This Program

This program can be used in a single-level classroom, multilevel classroom, homeschool, co-op group, or science club. Everything you need for a complete ocean study is included in this book. Intermediate students will need access to basic reference materials.

Take the time to read the entire *How to Use this Program* section and become familiar with the sections of this book described in the *Introduction*.

Begin a lesson by reading the *Teacher Pages* for that lesson. Choose the vocabulary words for each student and the activities to complete. Collect the materials you need for these activities.

Introduce each lesson with its corresponding *Lots of Science Library Book* by reading it aloud or asking a student to read it. (The *Lots of Science Library Books* are located after the *Teacher's Section* in this book.)

Discuss the concepts presented in the *Lots of Science Library Book*, focusing on the ones listed in your *Teacher's Section*.

Follow the directions for the activities you have chosen.

How to Use the Multilevel Approach

The lessons in this book include basic content appropriate for grades K–8 at different mastery levels. For example, throughout the teaching process, a first grader will be exposed to a lot of information but would not be expected to retain all of it. In the same lesson, a sixth-grade student will learn all the steps of the process, be able to communicate them in writing, and be able to apply that information to different situations.

In the *Lots of Science Library Books*, the words written in larger type are for all students. The words in smaller type are for upper level students and include more scientific details about the basic content, as well as interesting facts for older learners.

In the activity sections, icons are used to designate the levels of specific writing assignments.

> This icon ✎ indicates the Beginning level, which includes the nonreading or early reading student. This level applies mainly to kindergarten and first grade students.
>
> This icon ✎✎ is used for the Primary level. It includes the reading student who is still working to be a fluent reader. This level is designed primarily for second and third graders.
>
> This icon ✎✎✎ denotes the Intermediate level, or fluent reader. This level of activities will usually apply to fourth through eighth grade students.

If you are working with a student in seventh or eighth grade, we recommend using the assignments for the Intermediate level, plus at least one *Experiences, Investigations, and Research* activity per lesson.

No matter what grade level your students are working on, use a level of written work that is appropriate for their reading and writing abilities. It is good for students to review data they already know, learn new data and concepts, and be exposed to advanced information and processes.

Vocabulary Words

Each lesson lists vocabulary words that are used in the content of the lesson. Some of these words will be "too easy" for your students, some will be "too hard," and others will be "just right." The "too easy" words will be used automatically during independent writing assignments. Words that are "too hard" can be used during discussion times. Words that are "just right" can be studied by definition, usage, and spelling. Encourage your students to use these words in their own writing and speaking.

You can encourage beginning students to use their vocabulary words as you reinforce reading instruction and enhance discussions about the topic, and as words to be copied in cooperative, or teacher guided, writing.

Primary and Intermediate students can make a Vocabulary Book for new words. Instructions for making a Vocabulary Book are found on page 3. The Vocabulary Book will contain the word definitions and sentences composed by the student for each word. Students should also be expected to use their vocabulary words in discussions and independent writing assignments. A vocabulary word with an asterisk (*) next to it is designated for Intermediate students only.

Using 3D Graphic Organizers

The *3D Graphic Organizers* provide a format for students of all levels to conceptualize, analyze, review, and apply the concepts of the lesson. The *3D Graphic Organizers* take complicated information and break it down into visual parts so students can better understand the concepts. Most *3D Graphic Organizers* involve writing about the subject matter. Although the content for the levels will generally be the same, assignments and expectations for the levels will vary.

Beginning students may dictate or copy one or two "clue" words about the topic. These students will use the written clues to verbally communicate the science concept. The teacher should provide various ways for the students to restate the concept. This will reinforce the science concept and encourage the students in their reading and higher order thinking skills.

Primary students may write or copy one or two "clue" words and a sentence about the topic. The teacher should encourage students to use vocabulary words when writing these sentences. As students read their sentences and discuss them, they will reinforce the science concept while increasing their fluency in reading and higher order thinking skills.

Intermediate students may write several sentences or a paragraph about the topic. These students are also encouraged to use reference materials to expand their knowledge of the subject. As tasks are completed, students enhance their abilities to locate information, read for content, compose sentences and paragraphs, and increase vocabulary. Encourage these students to use the vocabulary words in a context that indicates understanding of the words' meanings.

Illustrations for the *3D Graphic Organizers* are found on the *Graphics Pages* and are labeled by the lesson number and a letter, such as 5–A. Your students may use these graphics to draw their own pictures, or cut out and glue them directly on their work.

Several of the *3D Graphic Organizers* will be used over a series of lessons. For this reason, you will need a storage system for each student's *3D Graphic Organizers*. A pocket folder or a reclosable plastic bag works well. See page 1 for more information on storing materials.

Investigative Loop™

The *Investigative Loop* is used throughout *Great Science Adventures* to ensure that your labs are effective and practical. Labs give students a context for the application of their science lessons so that they begin to take ownership of the concepts, increasing understanding as well as retention.

The *Investigative Loop* can be used in any lab. The steps are easy to follow, user friendly, and flexible.

Each *Investigative Loop* begins with a **Question or Concept.** If the lab is designed to answer a question, use a question in this phase. For example, the question could be: "How do Saturn and Earth compare in density?" Since the activity for this lab will show the density of two different objects, a question is the best way to begin this *Investigative Loop*.

If the lab is designed to demonstrate a concept, use a concept statement in this phase, such as: "The Moon reflects the light of the Sun." The lab will demonstrate that fact to the students.

After the **Question or Concept** is formulated, the next phase of the *Investigative Loop* is Research and/or Predictions. Research gives students a foundation for the lab. Having researched the question or concept, students enter the lab with a basis for understanding what they observe. Predictions are best used when the first phase is a question. Predictions can be in the form of a statement, a diagram, or a sequence of events.

 The **Procedure** for the lab follows. This is an explanation of how to set up the lab and any tasks involved in it. A list of materials for the lab may be included in this section or may precede the entire *Investigative Loop*.

Whether the lab is designed to answer a question or demonstrate a concept, the students' **Observations** are of prime importance. Tell the students what they are to focus upon in their observations. The Observation phase will continue until the lab ends.

 Once observations are made, students must **Record the Data**. Data may be recorded through diagrams or illustrations. Recording quantitative or qualitative observations of the lab is another important activity in this phase. Records may be kept daily for an extended lab or at the beginning and end for a short lab.

Conclusions and/or Applications are completed when the lab ends. Usually the data records will be reviewed before a conclusion can be drawn about the lab. Encourage the students to defend their conclusions by using the data records. Applications are made by using the conclusions to generalize to other situations or by stating how to use the information in daily life.

 Next, **Communicate the Conclusions**. This phase is an opportunity for students to be creative. Conclusions can be communicated through a graph, story, report, video, mock radio show, etc. Students may also participate in a group presentation.

Questions that are asked as the activity proceeds are called **Spark Questions**. Questions that the lab sparks in the minds of the students are important to discuss when the lab ends. The lab itself will answer many of these questions, while others may lead to a new *Investigative Loop*. Assign someone to keep a list of all Spark Questions.

 One lab naturally leads to another. This begins a new *Investigative Loop*. The phase called **New Loop** is a brainstorming time for narrowing the lab down to a new question or concept. When the new lab has been decided upon, the *Investigative Loop* begins again with a new Question or Concept.

Take the time to teach your students to make qualitative and quantitative observations. Qualitative observations involve recording the color, texture, shape, smell, size (such as small, medium, large), or any words that describe the qualities of an object. Quantitative observations involve using a standard unit of measurement to determine the length, width, weight, mass, or volume of an object.

All students will make a Lab Book, in the form of a Pocket Book, to record information about the *Investigative Loops*. Instructions are found on page 2. Your students will make a new Lab Book as needed to glue side–by–side to the previous one. Instructions can be found in the *Teacher's Section*.

Predictions, data, and conclusions about the *Investigative Loops* are written under the tabs of the Lab Book.

When you begin an *Investigative Loop*, ask your students to glue or draw the graphic of the experiment on the tab of the Lab Book. Each *Investigative Loop* is labeled with the lesson number and another number. These numbers are also found on the corresponding graphics.

During an *Investigative Loop*, beginning students should be encouraged to discuss their answers to all experiment questions. By discussing the topic, the students will not only learn the science concepts and procedures, but will be able to organize their thinking in a manner that will enhance their writing skills. This discussion time is very important for beginning students and should not be rushed.

After the discussion, work with the students to construct a sentence about the topic. Let them copy the sentence. Students can also write "clue" words to help them remember key points about the experiment and discuss it at a later time.

Primary students should be encouraged to verbalize their answers. By discussing the topic, students will learn the science concepts and procedures and learn to organize their thinking, increasing their ability to use higher–level thinking skills. After the discussion, students can complete the assignment using simple phrases or sentences. Encourage students to share the information they have learned with others, such as parents or friends. This will reinforce the content and skills covered in the lesson.

Even though Intermediate students can write the answers to the lab assignments, the discussion process is very important and should not be skipped. By discussing the experiments, students review the science concepts and procedures as well as organize their thinking for the writing assignments. This allows them to think and write at higher levels. These students should be encouraged to use their vocabulary words in their lab writing assignments.

Design Your Own Experiment

After an *Investigative Loop* is completed, intermediate students have the option to design their own experiments based on that lab. The following procedure should be used for those experiments.

Select a Topic based upon an experience in an *Investigative Loop*, science content, an observation, a high-interest topic, a controversial topic, or a current event.

Discuss the Topic as a class, in student groups, and with knowledgeable professionals.

Read and Research the Topic using the library, the Internet, and hands-on investigations and observations, when possible.

Select a Question that can be investigated and answered using easily obtained reference materials, specimens, and/or chemicals, and make sure that the question selected lends itself to scientific inquiry. Ask specific, focused questions instead of broad, unanswerable questions. Questions might ask "how" something responds, forms, influences, or behaves, or how it is similar or different to something else.

Predict the answer to your question, and be prepared to accept the fact that your prediction might be incorrect or only partially correct. Examine and record all evidence gathered during testing that both confirms and contradicts your prediction.

Design a Testing Procedure that gathers information that can be used to answer your question. Make sure your procedure results in empirical, or measurable, evidence. Don't forget to do the following:

> Determine where and how the tests will take place – in a natural (field work) or controlled (lab) setting.

> Collect and use tools to gather information and enhance observations. Make accurate measurements. Use calculators and computers when appropriate.

> Plan how to document the test procedure and how to communicate and display resulting data.

> Identify variables, or things that might prevent the experiment from being "fair." Before beginning, determine which variables have no effect, a slight effect, or a major effect on your experiment. Create a method for controlling these variables.

Conduct the Experiment carefully and record your findings.

Analyze the Question Again. Determine if the evidence obtained and the scientific explanations of the evidence are reasonable based upon what is known, what you have learned, and what scientists and specialists have reported.

Communicate Findings so that others can duplicate the experiment. Include all pertinent research, measurements, observations, controls, variables, graphs, tables, charts, and diagrams. Discuss observations and results with relevant people.

Reanalyze the Problem and if needed, redefine the problem and retest. Or, try to apply what was learned to similar problems and situations.

Experiences, Investigations, and Research

At the end of each lesson in the *Teacher's Section* is a category of activities entitled *Experiences, Investigations, and Research.* These activities expand upon concepts taught in the lesson, provide a foundation for further study of the content, or integrate the study with other disciplines. The following icons are used to identify the type of each activity.

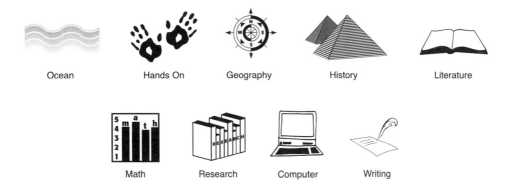

Cumulative Project

At the end of the program we recommend that students compile a Cumulative Project using the activities they have completed during their course of study. It may include the Investigative Loops, Lab Record Cards, and the *3D Graphic Organizers* on display.

Please do not overlook the Cumulative Project, as it provides immeasurable benefits for your students. Students will review all the content as they create the project. Each student will organize the material in his or her own unique way, thus providing an opportunity for authentic assessment and reinforcing the context in which it was learned. This project creates a format where students can make sense of the whole study in a way that cannot be accomplished otherwise.

These 3D Graphic Organizers are used throughout Great Science Adventures.

Fast Food and Fast Folds

"If making the manipulatives takes up too much of your instructional time, they are not worth doing. They have to be made quickly, and they can be, if the students know exactly what is expected of them. Hamburgers, Hot Dogs, Tacos, Mountains, Valleys, and Shutter–Folds can be produced by students, who in turn use these folds to make organizers and manipulatives." – Dinah Zike

Every fold has two parts. The outside edge formed by a fold is called the **"Mountain."** The inside of this edge is the **"Valley."**

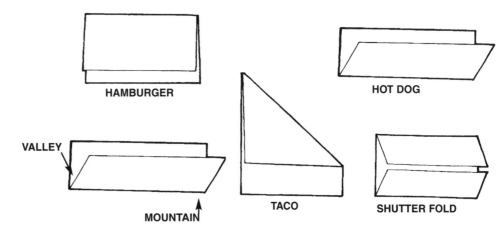

Storage – Book Bags

One–gallon reclosable plastic bags are ideal for storing ongoing projects and books that students are writing and researching.

Use strips of clear, 2" tape to secure 1" x 1" pieces of index card to the front and back of one of the top corners of a bag, under the closure. Punch a hole through the index cards. Use a giant notebook ring to keep several of the "Book Bags" together.

Label the bags by writing on them with a permanent marker.

Alternatively, the bags can be stored in a notebook if you place the 2" clear tape along the side of the storage bag and punch 3 holes in the tape.

Half Book

Fold a sheet of paper in half like a Hamburger.

Large Question and Answer Book

1. Fold a sheet of paper in half like a Hamburger. Fold it in half again like a Hamburger. Make a cut up the Valley of the inside fold, forming two tabs.

2. A larger book can be made by gluing Large Question and Answer Books "side–by–side."

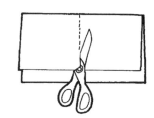

Pocket Book

1. Fold a sheet of paper in half like a Hamburger.

2. Open the folded paper and fold one of the long sides up two and a half-inch inches to form a pocket. Refold along the Hamburger fold so that the newly formed pockets are on the inside.

3. Glue the outer edges of the two and a half–inch fold with a small amount of glue.

4. Make a multi-paged booklet by gluing several Pocket Books "side–by–side."

5. Glue a construction paper cover around the multi–page pocket booklet.

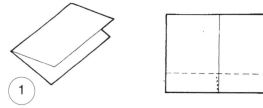

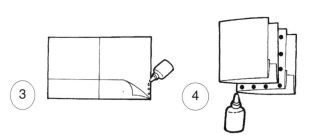

Side–by–Side

Some books can easily grow into larger books by gluing them side–by–side. Make two or more of these books. Be sure the books are closed, then glue the back cover of one book to the front cover of the next book. Continue in this manner, making the book as large as needed. Glue a cover over the whole book.

Vocabulary Book

1. Take two sheets of paper and fold each sheet like a Hot Dog.

2. Fold each Hot Dog in half like a Hamburger. Fold each Hamburger in half two more times and crease well. Unfold the sheets of paper, which are divided into sixteenths.

3. On one side only, cut the folds up to the Mountain top, forming eight tabs. Repeat this process on the second sheet of paper.

4. Take a sheet of construction paper and fold like a Hot Dog. Glue the back of one vocabulary sheet to one of the inside sections of the construction paper. Glue the second vocabulary sheet to the other side of the construction paper fold.

5. Vocabulary Books can be made larger by gluing them "side–by–side."

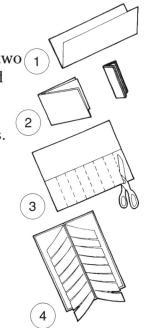

Layered Look Book

1. Stack two sheets of paper and place the back sheet one inch higher than the front sheet.

2. Bring the bottom of both sheets upward and align the edges so that all of the layers or tabs are the same distance apart.

3. When all tabs are an equal distance apart, fold the papers and crease well.

4. Open the papers and glue them together along the Valley/center fold.

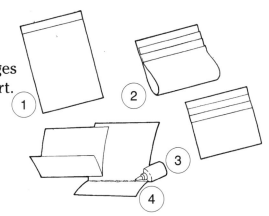

Matchbook

1. Fold a sheet of paper like a hamburger, but fold it so that one side is one inch longer than the other side.

2. Fold the one inch tab over the short side forming an envelope-like fold.

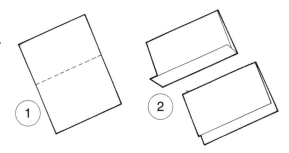

Great Science Adventures

Teacher's Section

Website addresses used as resources in this book are accurate and relevant at the time of publication. Due to the changing nature of the Internet, we encourage teachers to preview the websites prior to assigning them to students.

The authors and the publisher have made every reasonable effort to ensure that the experiments and activities in this book are safe when performed according to the book's instructions. We further recommend that students undertake these activities and experiments under the supervision of a teacher, parent, and/or guardian.

Great Science Adventures

Lesson 1

What do we know about Earth?

Ocean Concepts:
- Earth is one of the nine planets in our solar system and the third planet from the Sun.
- The Sun produces light and heat.
- One of Earth's unique qualities is the presence of water.
- About 75% of Earth's surface is covered with water.
- Continents separate the oceans: Pacific, Atlantic, Indian, Arctic, and Southern.
- Earth's crust is divided into continental plates which move apart or collide, or slide against each other.
- Molten rock can rise, cool, and form new rock on land or the ocean floor.

Vocabulary: Earth   Sun   solar system   continents   *continental plates   *Pangaea

Construct and Read: *Lots of Science Library Book #1.* (See page 65)

Activities:

Earth Graphic Organizer

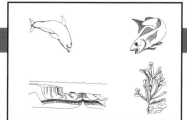

Focus Skills: mapping, labeling
Paper Handouts: a copy of Graphics 1A & B
Graphic Organizer: This is the beginning of a 22-Tab Graphic Organizer entitled *Discovering the Ocean* that will be used in this and future lessons. Title Graphic 1A and color the pictures for the cover page. Glue the cover on Graphic 1B at the glue line. On Graphic 1B:

✎ Copy or dictate the seven continents.

✎✎ List the seven continents. Using an atlas, list at least two cities or countries for each continent.

✎✎✎ List the seven continents. Using an atlas, list at least five cities or countries for each continent.

✎ Copy or dictate the five oceans. On the bottom section, draw a picture of the ocean.
Pacific, Atlantic, Indian, Arctic, Antarctic or Southern

✎✎   ✎✎✎   Write the five ocean names leaving room to add data from Lesson 3

Investigative Moving Continental Plates - Investigative Loop

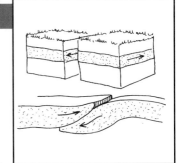

Focus Skill: observing a concept
Lab Materials: two popsicle sticks   pencil   dishwashing liquid   large, shallow pan   water
Paper Handouts: 8.5" x 11" sheet of paper   a copy of Lab Graphic 1-1   Lab Record Cards (index cards or pieces of 3" x 4" paper)
Graphic Organizer: Make a Pocket Book. See page 2 for directions. This is the student's Lab Book. Glue Lab Graphic 1-1 on the left pocket.
Concept: Continents drift apart.
Research: Read *Lots of Science Library Book #1*.
Procedure: Pour water into the pan until it is about half full. Gently place the popsicle sticks on the middle of the water's surface, leaving a small gap between the two popsicle sticks. Dip the pencil point in dishwashing liquid and insert the pencil tip between the two popsicle sticks.
Observations: Describe the motion of the popsicle sticks (which represent continental plates) after you inserted the pencil tip. **The popsicle sticks move like rafts on water.**
Record the Data: Label a Lab Record Card "Lab 1-2." Record your observations.
Conclusions: What can you conclude about the movement of continental plates?
Communicate the Conclusions: Label a Lab Record Card "Lab 1-2." Write your conclusions.
Spark Questions: Discuss questions sparked by this lab.
New Loop: Choose one question to investigate further.
Design Your Own Experiment: Select a topic based upon the experiences in the Investigative Loop. See page viii for more details.

Experiences, Investigations, and Research

Select one or more of the following activities for individual or group enrichment projects. Allow your students to determine the format in which they would like to report, share, or graphically present what they have discovered. This should be a creative investigation that utilizes your students' strengths.

1. Examine a globe, concentrating on the vastness of the ocean.

2. Use two stacks of bath towels to represent the layers of the Earth. Push the stacks together and observe what happens to the "layers." Pull the stacks apart and relate this to Earth's plate movement.

Great Science Adventures

Lesson 2

What do we know about water?

Ocean Concepts:
- The hydrosphere consists of all of the water on or near the Earth's surface.
- Water occurs in three states: solid, liquid, and gas.
- Water molecules stick together due to surface tension.
- Earth's water is constantly moving through the water cycle: evaporation, condensation, and precipitation.
- More than 97% of Earth's water is found in oceans.

Vocabulary: water   solid   liquid   gas   density   water cycle   *hydrosphere   *hydrogen   *oxygen   *surface tension   *evaporation   *condensation   *precipitation

Construct and Read: *Lots of Science Library Book #2.*

Activities:

Water - Graphic Organizer

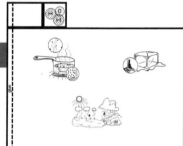

Focus Skill: charting
Paper Handouts: a copy of Graphic 2A
Graphic Organizer: Glue Graphic 2A under the previous page of *Discovering the Ocean* at the glue line. On the top of that page:
✎   Draw a picture of a water molecule.
✎✎ Draw a pie chart indicating the percentages of Earth's water found in the oceans, glaciers, groundwater, rivers, lakes, and streams, and the atmosphere: *ocean - 97%; Antarctica and glaciers - 2%; groundwater - .5%; rivers, lakes, streams - .02%; atmosphere - .0001%.*
✎✎✎ Complete ✎✎. Explain water's cohesive property and surface tension.

Under the picture of ice:
✎   Draw a picture of ice.
✎✎ Write clue words about water in its solid state: *molecules locked together in hexagonal crystals; molecules vibrate quickly.*
✎✎✎ Complete ✎✎. Explain why ice floats in water.

Under the pictures of water and water vapor:
- ✎ Draw a picture of water and water vapor and label accordingly.
- ✎✎ Write clue words about water in its liquid and gas states: liquid - *molecules move quickly so do not remain locked together; molecules move slowly enough to be attached to each other; water is less dense than in its solid state;* gas - *molecules so far apart and move so quickly that they rarely collide and do not become attached.*
- ✎✎✎ Explain the structure of water in its liquid and gas states.

Water Cycle - Graphic Organizer

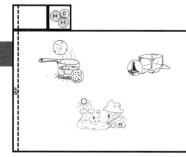

Focus Skill: explaining
Paper Handouts: *Discovering the Ocean Graphic Organizer*
Graphic Organizer: Label the water cycle graphic. Label: *Evaporation, Condensation,* and *Precipitation*.
- ✎ Color the picture.
- ✎✎ Write clue words about the water cycle: *heat from Sun turns water into water vapor;* condensation - *when water vapor meets cooler air, clouds form;* precipitation - *when clouds cannot hold any more water vapor, rain falls.*
- ✎✎✎ Explain the three stages of the water cycle.

Sticky Skin Water - Investigative Loop

Focus Skill: observing
Lab Materials: small hand mirror   water   bowl
Paper Handouts: Lab Book   Lab Record Card
   a copy of Lab Graphic 2-1
Graphic Organizer: Glue Lab Graphic 2-1 on the right pocket of the Lab Book.
Concept: Water has surface tension.
Research: Read *Lots of Science Library Book #2*.
Procedure: Place the mirror on a flat surface. Dip your finger in a bowl of water and gently flick a drop of water onto the mirror. Place another drop of water on top of the existing drop of water. Try changing the shape of the drop of water.
Observations: Describe the drop of water. **The drop of water is circular and dome-shaped**. Describe what happened after you added another drop of water. **The two drops of water stay together to form one drop**. Explain what happened when you changed the shape of the drop of water. **The water drops stayed together**.
Record the Data: On a Lab Record Card, sketch the lab. Show the different shapes you made with the drops of water.
Conclusions: Explain why the water molecules stick together. **Water molecules stick together by cohesion, which creates a "skin" on a drop of water. This effect is called surface tension.**
Communicate the Conclusions: On a Lab Record Card, record your conclusions.
Spark Questions: Discuss questions sparked by this activity.
New Loop: Choose one question and investigate it further.
Design Your Own Experiment: Select a topic based upon the experiences in the Investigative Loop. See page viii for more details.

Experiences, Investigations, and Research

Select one or more of the following activities for individual or group enrichment projects. Allow your students to determine the format in which they would like to report, share, or graphically present what they have discovered. This should be a creative investigation that utilizes your students' strengths.

1. How many drops of water will fit on a penny? Use an eyedropper to put drops on a penny. Estimate the number of drops and then determine the actual amount that will fit on the penny. Try the same procedure with a dime, nickel, and quarter.

2. Fill a glass with water. The water should be level with the top of the glass. Guess how many pennies you can slowly drop into the water before the water spills out of the glass. Use what you have learned in this lesson to explain your observations.

Notes

Great Science Adventures

Lesson 3

What do we know about the ocean?

Ocean Concepts:
- Oceans contain more than 97% of Earth's water.
- The Pacific, the largest and deepest ocean, holds about 50% of all Earth's water.
- Energy from the Sun keeps Earth's water in constant motion.
- Ocean water evaporates and eventually returns to Earth as rain or snow.
- Rainwater flows into rivers, picking up salts and eventually dumping them into the ocean.

Vocabulary: oceans   Pacific Ocean   Atlantic Ocean   Indian Ocean   Arctic Ocean   Southern Ocean   seas   rain   salty   *deltas

Construct and Read: *Lots of Science Library Book #3.*

Activities:

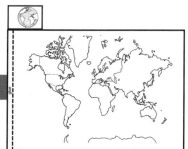

Oceans - Graphic Organizer

Focus Skill: describing, listing
Paper Handouts: *Discovering the Ocean* Graphic Organizer
Graphic Organizer: Beside each ocean name of 1B, write or dictate the area, and list the average and maximum depth:
   Pacific - *69,000,000 sq mi; average depth 14,000 ft; maximum depth 36,000 ft*
   Atlantic - *40,000,000 sq mi; average depth 11,000 ft; maximum depth 30,000 ft*
   Indian - *28,000,000 sq mi; average depth 13,000 ft; maximum depth 25,000 ft*
   Arctic - *5,440,000 sq mi; average depth 4,265 ft; maximum depth 17,880 ft*
   Southern or Antarctica - *11,000,000 sq mi; 185 million tons of water*

Experiences, Investigations, and Research

Select one or more of the following activities for individual or group enrichment projects. Allow your students to determine the format in which they would like to report, share, or graphically present what they have discovered. This should be a creative investigation that utilizes your students' strengths.

1. Make a Bar Graph indicating the average or maximum depth of each ocean area.

2. Make an ocean discovery time line that documents important events and accomplishments in oceanography.

3. Research how scientists map the ocean floor.

Great Science Adventures

Lesson 4

What is oceanography?

Ocean Concepts:
- The term oceanography originated from an expedition headed by C.W. Thomson and John Murray aboard the H.M.S. Challenger.
- Sonar helps to detect the location and distance of objects by the use of sound waves.
- Scuba divers can descend safely to only about 165 ft (50 m).
- Submersibles are used to descend into deeper waters.
- Research ships equipped with modern instruments help scientists study the ocean.

Vocabulary: oceanography   study   plants   animals   descend   *sonar *submersibles

Construct and Read: *Lots of Science Library Book #4.*

Activities:

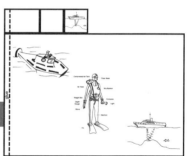

Oceanography - Graphic Organizer

Focus Skill: defining
Paper Handouts: *Discovering the Ocean* Graphic Organizer   a copy of Graphic 4A
Graphic Organizer: Glue Graphic 4A under the previous page of *Discovering the Ocean* at the glue line. On the top of that page:
✎   Color the pictures of a research ship, a scuba diver, or sonar.
✎✎ Write clue words about oceanography: *originated from Thomson's and Murray's expedition aboard the Challenger; includes study of ocean floor, currents, and temperature; chemical properties such as salinity; plants and animals; interaction with weather.*
✎✎✎ Complete ✎✎. Research one of the following studies aboard the H.M.S. Challenger, ALVIN, or Trieste Submarines.

Oceanography - Activity

Activity Materials: small plastic bottle   flexible drinking straw   dishpan   water
    modeling clay   2 pennies   tape   scissors or knife
Activity: Fill the dishpan with water. Using the point of a pair of scissors or a knife, make three holes on the side of the plastic bottle. Tape the pennies along the bottom of the bottle. Insert the short end of the straw in the bottle and secure it with modeling clay. Lower the bottle into the

dishpan of water, leaving the open end of the straw above the water. What happened? **The bottle (submarine) filled with water and sank to the bottom of the dishpan.** Now, place the straw in your mouth and blow. What happened? **The bottle rose to the surface.** Why? **The air forced the water out of the bottle, causing the submarine to rise to the surface.**

Experiences, Investigations, and Research

Select one or more of the following activities for individual or group enrichment projects. Allow your students to determine the format in which they would like to report, share, or graphically present what they have discovered. This should be a creative investigation that utilizes your students' strengths.

1. Google: ALVIN, HMS Challenger and Trieste Submarines. Report orally on your findings.

2. Compare and contrast the study of oceanography and marine biology. What are other fields of study that pertain to oceanography?

3. Why is the ocean referred to as the last frontier? Explain your answer by giving specific examples.

4. Write an outline of Jacques Cousteau's achievements, inventions, and expeditions. Orally present your investigation.

Great Science Adventures

Lesson 5

What is saltwater?

Ocean Concepts:
- The ocean contains salts including halite, or table salt.
- A gallon of seawater contains an average of six heaping tablespoons of salt.
- The Dead Sea is about seven times saltier than the oceans, so it contains only a few types of microorganisms.
- The density of seawater depends on its salinity and temperature.
- Salt remains behind during evaporation.
- Desalination is a process used to separate salt from saltwater.

Vocabulary: salt   minerals   float   salinity   microorganisms   *halite
*sodium chloride   *calcium sulfate   *calcium carbonate   *desalinization

Construct and Read: *Lots of Science Library Book #5.*

Activities:

Saltwater - Graphic Organizer

Focus Skill: explaining
Paper Handouts: *Discovering the Ocean* Graphic Organizer
Graphic Organizer: On the bottom of Graphic 5A:
- Draw a picture of yourself floating easily in the salty Dead Sea.
- Write clue words about saltwater: *average of six heaping tablespoons of salt in a gallon of seawater; sea animals use calcium sulfate and calcium carbonate to produce their shells and bones; some bodies of water are more salty than others; the greater the salinity, the denser the water; density of seawater depends on salinity and temperature; when evaporation occurs, salt remains.*
- Complete ✎✎. Explain the process of desalination.

Investigative Loop - Desalinization

Focus Skill: following a procedure
Lab Materials: 4 cups water   1/2 cup salt   pot with lid
Paper Handouts: 8.5" x 11" sheet of paper   Lab Book
    Lab Record Cards   a copy of Lab Graphic 5-1
Graphic Organizer: Make a Pocket Book and glue Lab Graphic 5-1 on the left pocket. Glue the Pocket Book side-by-side to the Lab Book.

Concept: Freshwater can be taken from saltwater through desalination.
Research: Read *Lots of Science Library Book #5*.
Procedure: Mix the water and salt in the pot and heat over a stove. Let it come to a boil and simmer with the lid on the pot. Every few minutes, lift the lid and pour the water from the lid into a cup. Repeat the procedure until enough water fills the cup to taste it.
Observations: Taste the condensed water in the cup. Describe it. Is it salty?
Record the Data: Label a Lab Record Card "Lab 5-1." Record your observations.
Conclusions: Draw conclusions about desalination.
Communicate the Conclusions: Label a Lab Record Card "Lab 5-1." Write your conclusions.
Spark Questions: Discuss questions sparked by this lab.
New Loop: Choose one question to investigate further.
Design Your Own Experiment: Select a topic based upon the experiences in the Investigative Loop. See page viii for more details.

Investigative Loop - Water Density

Focus Skill: observing a concept
Lab Materials: water   3/4 cup salt   2 crayons   2 glass containers
Paper Handouts: Lab Book   Lab Record Cards
   a copy of Lab Graphic 5-2
Graphic Organizer: Glue Lab Graphic 5-2 on the right pocket of the Lab Book.
Concept: Salt in ocean water increases the water's density.
Research: Read *Lots of Science Library Book #5*.
Procedure: Pour two cups of water into each glass container. In one container, add the salt and stir until dissolved. Carefully place a crayon in the fresh water. Place the other crayon in the saltwater.
Observations: What happened to the crayon in the freshwater? What happened to the crayon in the saltwater? **The crayon in the freshwater sank; the crayon in the saltwater floated.**
Record the Data: Label a Lab Record Card "Lab 5-2." Record your observations.
Conclusions: What can you conclude about the density of water? **Saltwater is more dense than freshwater, so it is able to keep the crayon afloat.**
Communicate the Conclusions: Label a Lab Record Card "Lab 5-2." Write your conclusions.
Spark Questions: Discuss questions sparked by this lab.
New Loop: Choose one question to investigate further.
Design Your Own Experiment: Select a topic based upon the experiences in the Investigative Loop. See page viii for more details.

Experiences, Investigations, and Research

Select one or more of the following activities for individual or group enrichment projects. Allow your students to determine the format in which they would like to report, share, or graphically present what they have discovered. This should be a creative investigation that utilizes your students' strengths.

1. Google: desalination

2. Research what would happen to life in the ocean if the ice on land melted and eventually mixed with oceanwater.

3. How do marine animals depend upon salt water for survival? Could freshwater animals live in salt water? Why or why not?

4. Investigate what happens when less dense freshwater from rivers empty into the ocean. For example, what happens at the mouth of the Amazon River?

Notes

Great Science Adventures

Lesson 6

What do we know about the ocean floor?

Ocean Concepts:
- The slope of the ocean floor from the continents is called the continental shelf. It slopes down to about 650 feet (200 m).
- At the break of the shelf, the ground takes a steep downward slope to the depths of the ocean floor; this is called the continental slope.
- The abyssal plain covers about one-half of the deep ocean floor.
- Beyond the plains are mountain ranges and deep trenches.
- Hydrothermal vents blasting out sulfur-rich hot water were recently found on the ocean floor by scientists.

Vocabulary: slope   downward   shallow   valleys   trenches   mountains   steep
*continental shelf   *continental slope   *abyssal plains   * abyssal hills   *oceanic ridge

Construct and Read: *Lots of Science Library Book #6.*

Activities:

Ocean Floor - Graphic Organizer

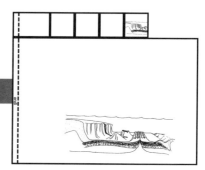

Focus Skill: mapping
Paper Handouts: *Discovering the Ocean* Graphic Organizer
a copy of Graphic 6A
Graphic Organizer: Glue Graphic 6A under the previous page of *Discovering the Ocean* at the glue line. Label each section of the Graphic: *Continental Shelf, Continental Slope, Abyssal Plains,* and *Mountains and Trenches*. Next to each label:

Continental Shelf:
✎ Draw pictures of plants and animals living in these waters.
✎✎ Write clue words about the continental shelf: *sloping gradually to about 650 ft (200 m); sunlit zone makes up about 10% of ocean; important because plants and animals thrive here; sun filters through the water to give light and warmth; almost all world's seafood comes from here; all minerals and petroleum removed from oceans are extracted here.*
✎✎✎ Complete ✎✎. Research the fishing industry.

Continental Slope:
✎ Draw a picture of the continental slope.
✎✎ Write clue words about the continental slope: *continental shelf drops steeply into the continental slope.*

✎✎✎ Describe the continental slope.

Abyssal Plains:
✎ Draw a picture of the abyssal plains.
✎✎ Write clue words about the abyssal plains: *at the end of the continental slope; covered with mud; water is about 6,500-13,000 ft (2,000-4,000 m) deep; water is black and cold.*
✎✎✎ Describe the abyssal plains.

Mountains and Trenches:
✎ Draw a picture of tall mountains and deep trenches under the ocean.
✎✎ Write clue words about the hills, mountains, and trenches that lie beyond the abyssal plains: *mountains can be twice the height of Mt. Everest, which is 29,000 ft (9,000 m); trenches can be six times deeper than Grand Canyon, 35,800 ft (11,000 m), and can stretch about 1,550 mi (2,500 km) long.*
✎✎✎ Complete ✎✎. Describe oceanic ridges.

Experiences, Investigations, and Research

Select one or more of the following activities for individual or group enrichment projects. Allow your students to determine the format in which they would like to report, share, or graphically present what they have discovered. This should be a creative investigation that utilizes your students' strengths.

1. Google: ocean floor

2. Compare and contrast the abyssal plains to a flat region on Earth's surface.

3. Explain why the tallest mountain on Earth is found in the Pacific Ocean.

Great Science Adventures

Lesson 7

What are the ocean layers?

Ocean Concepts:
- Ocean waters are divided into five layers: Sunlit Zone, Twilight Zone, Dark Zone, Abyss, and Trenches.
- The Sunlit Zone is 600 feet (183 m) deep, has warm water, and contains most of the plants and animals in the ocean.
- The Twilight Zone is the next 2,700 feet (823 m) with temperatures as low as 41 degrees F (5 degrees C).
- Extending to 13,200 feet (4,025 m) below the ocean surface is the Dark Zone. The only visible light is produced by creatures in that zone.
- The Abyss extends from 13,200 feet (4,025 m) to 19,800 feet (6,000 m) below the ocean's surface, with near freezing water and no light at all.
- Below 19,800 feet (6,000 m) lie the lower areas of trenches and canyons. Even with the freezing temperatures and blackness, life can be found.

Vocabulary: temperature   depth   radiates   *Sunlit Zone   *Twilight Zone
*Dark Zone   *Abyss   *Trenches

Construct and Read: *Lots of Science Library Book #7.*

Activities:

Ocean Layers - Graphic Organizer

Focus Skill: labeling
Paper Handouts: *Discovering the Ocean* Graphic Organizer   a copy of Graphic 7A
Graphic Organizer: Glue Graphic 7A under the previous page of *Discovering the Ocean* at the glue line. Label each section of the Graphic: *Sunlit Zone, Twilight Zone, Dark Zone, Abyss,* and *Trenches and Canyons*. On the left of each label:

Sunlit Zone section:
- Draw pictures of plants and animals that thrive in the sunlit zone.
- Write clue words about the sunlit zone. Include examples of plants and animals that thrive in this zone: *the top 600 ft (183 m) receives sunlight; wide variety of plants and animals; near shore are seaweed and other plants that provide food and shelter; in deeper open seas are whales, phytoplankton, and zooplankton.*
- Describe the conditions in the Sunlit Zone.

Twilight Zone:
- Draw pictures of animals that live in the twilight zone.
- Write clue words about the twilight zone. Include examples of animals that live in this zone: *next 3,000 ft (914 m); little sunlight; many animals glow in dark; light-producing bacteria help some fish with defense or to find a mate or prey; most animals are scavengers; lanternfish; anglerfish; hatchetfish.*
- Describe the conditions in the Twilight Zone.

Dark Zone:
- Draw pictures of animals that live in the dark zone.
- Write clue words about the dark zone. Include examples of animals that live in this zone: *no sunlight; 3,280 - 20,000 ft (1,000 - 6,096 m); bottom of abyssal plain is completely dark and very cold; unusual-looking animals withstand great pressure; gulper eel.*
- Describe the condition of the Dark Zone.

Abyss:
- Color the pictures of the Abyss.
- Write clue words about the Abyss: *extends 13,200 ft - 19,800 ft. below surface, near freezing, no light, mud is thick.*
- Describe the condition of the Abyss.

Trenches and Canyons:
- Color the pictures of trenches and canyons.
- Write clue words about trenches and canyons: *beyond 19,800 ft. below the Abyss, black, near freezing.*
- Describe the conditions in the trenches and canyons.

Sunlit Zone and Dark Zone - Activity

Activity Materials: outdoor area with grass   bowl

Activity: Find an area of green grass. Place a bowl upside-down in one section and leave it undisturbed for about one week. At the end of the week, remove the bowl and compare the grass under the bowl to the grass around the bowl. What did you observe? How does this demonstrate life in the sunlit zone vs. life in the dark zone? **The grass under the bowl turned yellow; the surrounding grass remained green. Plants need light to carry on photosynthesis, or produce their own food. Chlorophyll, the substance that gives plants their green color, is a necessary substance in photosynthesis. Marine plants, like all green plants, need sunlight to produce their own food. Green plants are more abundant in the sunlit zone; plant life decreases deeper in the ocean. The dark zone receives very little to no sunlight.**

Water Pressure in the Deep - Investigative Loop

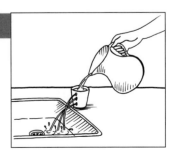

Focus Skill: demonstrating a concept
Lab Materials: large plastic or styrofoam drinking cup   pencil
    pitcher of water   dish pan or sink   masking tape   a partner
Paper Handouts: 8.5" x 11" sheet of paper   Lab Book
    Lab Record Cards   a copy of Lab Graphic 7-1

Graphic Organizer: Make a Pocket Book and glue Lab Graphic 7-1 on the left pocket. Glue the Pocket Book side-by-side to the Lab Book.
Concept: Water pressure increases in deeper waters.
Research: Read *Lots of Science Library Book #7*.
Procedure: Using the point of a pencil, make three holes on a slight slant, beneath each other, on the side of a cup. Cover the holes with masking tape. Place the cup by the dish pan or sink. Pour water into the cup. Remove the tape and ask your partner to continue pouring water.
Observations: Describe the water flowing from each hole. The distance that the water has to flow is less from the top hole than from the bottom hole.
Record the Data: On a Lab Record Card, sketch the lab and indicate the flow of water with arrows.
Conclusions: How does the lab show that water pressure increases in deeper waters? **The distance of the flow of water from the bottom hole is greatest.**
Communicate the Conclusions: On a Lab Record Card, record your conclusions.
Spark Questions: Discuss questions sparked by this activity.
New Loop: Choose one question and investigate it further.
Design Your Own Experiment: Select a topic based upon the experiences in the Investigative Loop. See page viii for more details.

Experiences, Investigations, and Research

Select one or more of the following activities for individual or group enrichment projects. Allow your students to determine the format in which they would like to report, share, or graphically present what they have discovered. This should be a creative investigation that utilizes your students' strengths.

1. Google: ocean zones

2. Research the importance of the first 6" of the ocean's surface.

3. Create a Pie Graph representing the depths of the ocean levels.

Notes

Great Science Adventures

Lesson 8

What lives in the ocean layers?

Ocean Concepts:
- Two types of plants are found in the ocean, those with roots attached to the ocean bottom and those which drift in the water.
- Rooted plants are only capable of growing in the Sunlit Zone.
- Phytoplankton makes up about 99% of plant life in the ocean.
- Marine animals are surface drifters, free swimmers, or bottom dwellers.
- Surface drifters are unable to propel themselves; they move by the wind or current.
- Free swimmers can propel themselves through the water.
- Bottom dwellers live on the ocean floor.
- In the Twilight Zone and below, there is no sunlight so plants cannot grow. Animals in these areas sometimes produce their own light.

Vocabulary: surface dweller   free swimmer   bottom dweller   *phytoplankton

Construct and Read: *Lots of Science Library Book #8.*

Activities:

Ocean Layers - Graphic Organizer

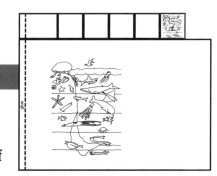

Focus Skill: acquiring information
Paper Handouts: *Discovering the Ocean* Graphic Organizer
Graphic Organizer: Inside each layer draw pictures of the types of sea life in that layer.

✎ Draw an animal or plant from each ocean layer.
✎✎ Write clue words about the types of sea life: *plants with roots to ocean bottom in Sunlit Zone; single-celled floating plants called phytoplankton; surface dwelling animals cannot propel themselves, free swimmers propel themselves, bottom dwellers live on ocean floor, no plants in Twilight Zone and below, few animals, most have own lights.*
✎✎✎ Describe sea life in the different ocean layers.

Fish Print - Activity

Activity Materials: tempera paint or poster paint   fresh, whole fish (with head, fins, and scales)   paintbrush   sheet of paper   newspaper   paper towels

Activity: Wash the fish and dry well with paper towel. Place the fish on newspaper. Spread out the fins. Brush paint on one side of the fish. Carefully place a sheet of paper on top of the fish. Remove the paper without making smudges. Gyotaku (gyo-TA-koo), the art of fish printing, originated in Japan in the mid to late 19th century.

Experiences, Investigations, and Research

Select one or more of the following activities for individual or group enrichment projects. Allow your students to determine the format in which they would like to report, share, or graphically present what they have discovered. This should be a creative investigation that utilizes your students' strengths.

1. Draw a Venn Diagram to compare animals of the Sunlit Zone and animals of the Dark Zone.

2. Write a first person story of an animal from each ocean layer.

Great Science Adventures

Lesson 9

What are waves?

Ocean Concepts:
- Waves, the continuously moving ridges we see on ocean water, are caused by winds.
- The size and speed of waves depend on the speed of the wind, how long it blows, and the fetch.
- In the open seas, waves appear in different sizes and shapes and travel in different directions.
- Strong, steady winds that blow across the ocean eventually produce swells.
- Waves lift water particles up and down in a little circle; only the energy of the waves moves forward.
- Ocean waves move in transverse waves.
- The top of a wave is called the crest; the bottom is called the trough.
- The distance between two crests is called the wavelength; the distance from the crest to the trough is the wave's height.

Vocabulary: moving   waves   winds   speed   forward   shape   *fetch
*transverse waves   *trough   *crest   *breaker   *tsunami

Construct and Read: *Lots of Science Library Book #9.*

Activities:

Waves - Graphic Organizer

Focus Skill: identifying
Paper Handouts: *Discovering the Ocean* Graphic Organizer   a copy of Graphics 9A
Graphic Organizer: Glue Graphic 9A under the previous page of *Discovering the Ocean* at the glue line.

✎ Color the waves in the manner in which they move across the ocean.
✎✎ Write clue words about how the energy of ocean waves moves in transverse waves: *ocean water does not move with the wave; waves lift water particles up and down in a little circle; a floating bottle does not move along with the wave; only the energy of the waves moves forward; water in waves moves perpendicular to wave direction; water moves up and down as waves move toward shore; size and speed of waves vary depending on wind speed, how long it blows, and fetch, which is distance a wave travels; faster and longer wind blows, the greater the fetch; greater the fetch, bigger the waves; strong, steady winds blow across ocean and eventually produce swells.*

✏️✏️✏️ Explain transverse waves. Compare ocean waves to light and sound waves. Label the parts of a wave. On the bottom section, explain how waves form and describe the conditions which increase wave size.

Wave Friends

Teacher's Note: This activity requires about ten students.
Activity: Stand nine students side-by-side and ask them to interlock their arms. Ask the tenth student to stand at the end of the line and gently pull the end student forward and then after a few seconds, pull the student back. What did you observe? What happened to the line of students when the end student was pulled forward and back? Where did the energy come from that made the line of students move? **The energy came from the tenth student.** What kind of wave was produced? **Transverse wave.**

Jump Rope Waves

Activity Materials: jump rope or long rope
Activity: Tie one end of the jump rope to a door or ask a partner to hold it. Flick the rope up and down. How does this activity show how waves move? **The rope moves up and down as the waves move forward; however, the rope does not move forward with the waves. Ocean waves move in transverse waves. They carry water particles up and down in a circular manner.**

Waves in Action - Investigative Loop

Focus Skill: demonstrating a concept.
Lab Materials: dish pan   a piece of board to fit the dish pan, as shown   soil   sand   water
Paper Handouts: 8.5" x 11" sheet of paper   Lab Book   Lab Record Cards
     a copy of Lab Graphic 9-1
Graphic Organizer: Glue Lab Graphic 9-1 on the right pocket of the Lab Book.
Concept: Waves can change the shape of a beach.
Research: Read *Lots of Science Library Book #9*.
Procedure: Wedge a piece of board in the dish pan, dividing it into about 2/3 and 1/3. Place soil in the 1/3 section and soak it with water so it is mud-like. Pack the mud down with your hands and let it dry about 3 days. After the mud has dried, pour about 2 inches of sand on the other side of the board. Pour water on the sand until it reaches about 2 inches above sand level. Carefully remove the board. Make waves by moving the board back and forth in the water. Try small waves and large waves.

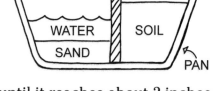

Observations: What did you observe? **The waves eroded the soil (beach).**
Record the Data: Label a Lab Record Card "Lab 9-1." Describe your observations or draw a diagram of the lab.
Conclusions: Explain how waves can change the shape of a beach.
Communicate the Conclusions: On a Lab Record Card, record your conclusions.
Spark Questions: Discuss questions sparked by this activity.
New Loop: Choose one question and investigate it further.
Design Your Own Experiment: Select a topic based upon the experiences in the Investigative Loop. See page viii for more details.

Bobbing Cork - Activity

Activity Materials: cork or any small, buoyant object   bathtub or large dishpan
flat piece of board

Activity: Place the cork or other object on the water. Make waves using the board, and observe the cork. Did the cork move with the wave? Explain. **The cork moved up and down with the wave but did not move across with the wave. Waves lift the cork up and down in a little circle; only the energy of the waves moves forward.**

Experiences, Investigations, and Research

Select one or more of the following activities for individual or group enrichment projects. Allow your students to determine the format in which they would like to report, share, or graphically present what they have discovered. This should be a creative investigation that utilizes your students' strengths.

1. Google: ocean waves

2. Make a Time Line on the history of surfing.

3. Investigate and map the wave size at different locations in the world.

4. Investigate the causes and effects of the 2004 tsunami in Asia.

Notes

Great Science Adventures

Lesson 10

What are tides?

Ocean Concepts:
- Tides are the daily rise and fall of water.
- Tides are caused by the gravitational pull of the moon and also of the Sun.
- The waters facing the moon are pulled toward it, causing high tide.
- Most places along the coast have two high and two low tides daily.
- Spring tides, the biggest tides, occur during full moons and new moons when the Earth lines up with the moon and the Sun.

Vocabulary: high   far   gravity   daily   time   size   high tide   low tide   Sun   moon   *spring tide   *neap tide

Construct and Read: *Lots of Science Library Book #10.*

Activities:

Tides - Graphic Organizer

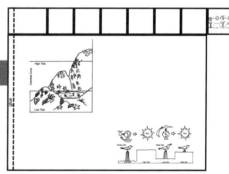

Focus Skill: communicating
Paper Handouts: *Discovering the Ocean* Graphic Organizer
a copy of Graphic 10A
Graphic Organizer: Glue Graphic 10A under the previous page of *Discovering the Ocean* at the glue line. On the top of that page:

✎ Draw pictures of the ocean during high tide and low tide.
✎✎ Write clue words about tides: *daily rise and fall of water; caused by gravitational pull of moon and Sun; as Earth rotates around Sun, waters pulled towards the moon, causing high tide; most places on coast have two high and two low tides daily; time and size of tides can vary due to type of coastline; wider coastlines have slight changes; narrow bays have greater tidal range.*
✎✎✎ Define tides and explain the cause of high and low tides. Research the tidal range of the Bay of Fundy and its fishing history.

On the bottom of the Graphic:
✎ Color the picture of a high spring tide with a full moon.
Color the picture of a neap tide with only a partial moon.
✎✎ Write clue words about each: Spring - *twice a month when Earth lines up with moon and*

Sun; gravitational pull from moon and Sun produce biggest tides; occurs during full moons and new moons. Neap - twice a month when moon and Sun pull at right angles to each other; produces smallest tides because gravitational pull of moon and Sun cancel each other out.

✎✎✎ Explain the *Spring* and *Neap Tides* and when each occurs. Research Newton's theory of gravity, which helped to further explain the effect of the moon and the Sun on tides. Write a summary to the left of Graphic 10A on the back of Graphic 9A.

Experiences, Investigations, and Research

Select one or more of the following activities for individual or group enrichment projects. Allow your students to determine the format in which they would like to report, share, or graphically present what they have discovered. This should be a creative investigation that utilizes your students' strengths.

1. Google: ocean tides

2. Research the Bay of Fundy and the world's highest tides. Determine why the tides are so high in this geographic location.

3. Explain why fishermen are interested in the times of high and low tides. What other professions might need this information?

Great Science Adventures

Lesson 11

What are ocean currents?

Ocean Concepts:
- Currents are underwater streams that move through the oceans.
- Surface currents are caused by the Earth's rotation, winds, and land masses.
- Deep currents, caused by temperature differences and salinity, mix warm and cold waters.
- The ocean's salinity and temperature cause currents to flow up and down.

Vocabulary: currents   surface   rotation   winds   land masses   saltiness

Construct and Read: *Lots of Science Library Book #11.*

Activities:

Ocean Currents - Graphic Organizer

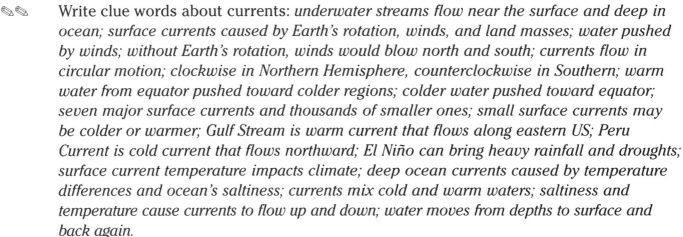

Focus Skill: defining
Paper Handouts: *Discovering the Ocean* Graphic Organizer
a copy of Graphic 11A
Graphic Organizer: Glue Graphic 11A under the previous page of *Discovering the Ocean* at the glue line. On the top of that page:

✏️ Trace the ocean currents.

✏️✏️ Write clue words about currents: *underwater streams flow near the surface and deep in ocean; surface currents caused by Earth's rotation, winds, and land masses; water pushed by winds; without Earth's rotation, winds would blow north and south; currents flow in circular motion; clockwise in Northern Hemisphere, counterclockwise in Southern; warm water from equator pushed toward colder regions; colder water pushed toward equator; seven major surface currents and thousands of smaller ones; small surface currents may be colder or warmer; Gulf Stream is warm current that flows along eastern US; Peru Current is cold current that flows northward; El Niño can bring heavy rainfall and droughts; surface current temperature impacts climate; deep ocean currents caused by temperature differences and ocean's saltiness; currents mix cold and warm waters; saltiness and temperature cause currents to flow up and down; water moves from depths to surface and back again.*

✏️✏️✏️ Define surface currents and deep ocean currents. Give examples and their movements.

Cold and Warm Water - Investigative Loop

Focus Skill: observing
Lab Materials: two identical glass jars   index cards   water
   red food coloring   dishpan
Paper Handouts: 8.5" x 11" sheet of paper   Lab Book
   Lab Record Cards   a copy of Lab Graphic 11-1
Graphic Organizer: Make a Pocket Book and glue Lab Graphic 11-1 on the left pocket. Glue the Pocket Book side-by-side to the Lab Book.
Concept: Warm water is less dense than cold water.
Research: Read *Lots of Science Library Book #11*.
Procedure: Place one jar in the dishpan and fill the jar with very warm water. Add a couple of drops of food coloring to the jar of warm water. Fill the other jar with cold water and cover it with an index card. Holding the index card in place, turn over the jar of cold water and gently place it on top of the jar with warm water. (Air pressure will keep the index card in place.) Carefully slide the index card out.
Observations: What do you observe? **The colored water rises up into the cold water.**
Record the Data: On a Lab Record Card, sketch the lab. Draw arrows showing the warm water rising.
Conclusions: Explain how the lab shows that warm water is less dense than cold water.
Communicate the Conclusions: On a Lab Record Card, write your conclusions.
Spark Questions: Discuss questions sparked by this activity.
New Loop: Choose one question and investigate it further.
Design Your Own Experiment: Select a topic based upon the experiences in the Investigative Loop. See page viii for more details.

Experiences, Investigations, and Research

Select one or more of the following activities for individual or group enrichment projects. Allow your students to determine the format in which they would like to report, share, or graphically present what they have discovered. This should be a creative investigation that utilizes your students' strengths.

1. Google: ocean currents

2. Google: www.geophys.washington.edu/tsunami

3. Google: www.globalchange.com/tsunami.htm

4. Describe the Gulf Stream and explain its importance. Color the Gulf Stream on a map and explain how it changes.

Great Science Adventures

Lesson 12

How does the ocean affect weather?

Ocean Concepts:
- Oceans play a key role in Earth's atmosphere and weather.
- Ocean temperature changes more slowly than land temperature.
- Tropical oceans hold in heat, while polar waters remain cold all year.
- The consistent temperatures of the oceans help moderate land temperatures.
- Water evaporates, condenses, and returns to land as rain or snow.
- Storms and hurricanes begin in warm, tropical seas.
- Earthquakes under the ocean can cause tsunamis.
- Storm surges occur when waters rise underneath hurricanes.

Vocabulary: weather   hot   cold   hurricane   tsunamis   storm surges

Construct and Read: *Lots of Science Library Book #12.*

Activities:

Oceans Affect Weather - Graphic Organizer

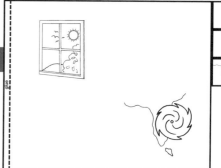

Focus Skill: explaining a relationship
Paper Handouts: *Discovering the Ocean* Graphic Organizer
   a copy of Graphic 12A
Graphic Organizer: Glue Graphic 12A under the previous
   page of *Discovering the Ocean* at the glue line. On the top of that page:
   ✎ Draw a picture of a hurricane over the ocean.
   ✎✎ Write clue words about the interaction between oceans and weather: *ocean temperatures change slowly; land masses may be very hot or very cold; tropical oceans hold in heat while polar waters remain cold all year; land quickly warms up or cools with seasons; consistent ocean temperatures help moderate the temperatures on land by warming and cooling air masses that move over the ocean surface; ocean's effect is more obvious in coastal areas but influences all areas of land; Earth maintains very consistent temperature due to oceans; heat from Sun evaporates water; water vapor meets cooler air and condenses; when cool air cannot hold any more water, rain falls; trade winds are more regular over oceans than land, and as they move over tropical waters, they pick up moisture and bring heavy rain to mountainous areas; storms and hurricanes begin in warm, tropical seas and weaken as they move over land.*
   ✎✎✎ Explain the interaction between oceans and weather. Compare and contrast tsunamis with storm surges.

Holding Heat - Activity

Teacher's Note: The purpose of this activity is to introduce the student to the concept of water's high specific heat. A true scientific experiment would include more precise variables.

Activity Materials: 2 small styrofoam cups   clean sand (Purchase the sand rather than using sand from outside so it will not contain moisture.)   2 thermometers   lamp

Activity: Pour sand in one cup and water in another cup. Leave them overnight to achieve room temperature. Insert a thermometer into the middle of each cup and take a reading after a few minutes. Re-insert the thermometers and place the lamp over the cups. Place the lamp carefully so both cups receive the same amount of heat. Take a reading four times at 15-minute intervals. What can you conclude?

Rain - Investigative Loop

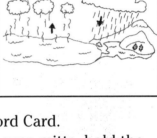

Focus Skill: predicting an outcome
Lab Materials: kettle of water   stove   pot   oven mitts   ice cubes
Paper Handouts: Lab Book   Lab Record Card   a copy of Graphic 12-1
Graphic Organizer: Glue Graphic 12-1 on the right pocket of the Lab Book.
Question: How does rain form?
Research: Read *Lots of Science Library Book #12*.
Predictions: Predict how rain forms. Write your prediction on a Lab Record Card.
Procedure: Boil a kettle of water. Place several ice cubes in a pot. Using oven mitts, hold the pot over the kettle.
Observations: Observe the bottom of the pot. What do you observe? **Water droplets form on the bottom of the pot and eventually fall as "rain."**
Record the Data: Label a Lab Record Card "Lab12-1." Draw a picture of the lab.
Conclusions: Explain how oceans affect weather. **Heat from the Sun causes water in the oceans, lakes, rivers, and ground to evaporate and become water vapor. Water vapor rises into the atmosphere. As water vapor meets cooler air, condensation occurs. The cool air cannot hold the water vapor, and water vapor turns back into water droplets as precipitation.**
Communicate the Conclusions: On a Lab Record Card, compare your observations and conclusions with your predictions.
Spark Questions: Discuss questions sparked by this lab.
New Loop: Choose one question to investigate further.
Design Your Own Experiment: Select a topic based upon the experiences in the Investigative Loop. See page viii for more details.

Experiences, Investigations, and Research

Select one or more of the following activities for individual or group enrichment projects. Allow your students to determine the format in which they would like to report, share, or graphically present what they have discovered. This should be a creative investigation that utilizes your students' strengths.

1. Investigate El Niño and La Niña. Explain their impact on weather.

Great Science Adventures

Lesson 13

What is the intertidal zone and what lives there?

Ocean Concepts:
- The rise and fall of tides create the intertidal zone.
- During high tide, the intertidal zone is covered with water.
- The intertidal zone is made up of the lower, middle, upper, and splash zones.
- The lower intertidal zone is almost always covered with water.
- The middle intertidal zone is host to organisms that can survive underwater during high tide and exposure to air during low tide, such as clams, shellfish, crabs, sea anemones, and mussels.
- The upper intertidal zone, along with the splash zone, receives the least amount of water exposure so plants and animals come into contact with water only during the highest tides and when waves splash. Animals that survive here include periwinkles, barnacles, and snails.
- Ebb tides leave behind tidal pools, which provide a habitat for many plants and animals.

Vocabulary: lower   middle   upper   splash   predator   prey   air   exposure   water   *ebb tide   *intertidal zone   *tidal pools

Construct and Read: *Lots of Science Library Book #13.*

Activities:

Intertidal Zones - Graphic Organizer

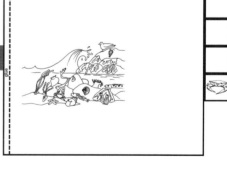

Focus Skill: describing
Paper Handouts: *Discovering the Ocean* Graphic Organizer
    a copy of Graphics 13A
Graphic Organizer: Glue Graphic 13A under the previous page of *Discovering the Ocean* at the glue line. Label each section of the intertidal zone: *Upper Intertidal Zone, Middle Intertidal Zone,* and *Lower Intertidal Zone with Tide Pools.* Next to the Graphic on this page:

Upper Intertidal Zone:
- Draw a picture of animals living in the *Splash Zone* and the *Intertidal Zone*.
- Write clue words about the conditions in the *Splash Zone* and the *Upper Intertidal Zone* and include examples of organisms living here: *receives least amount of water exposure; organisms come into contact with water only during highest tides and when*

waves splash; organisms must withstand long exposure to Sun without moisture; least populated zone; periwinkles, barnacles, and snails.

- Describe conditions in the *Splash Zone* and the *Upper Intertidal Zone*. Choose one organism that lives here and write a short paragraph about it.

Middle Intertidal Zone:
- Draw a picture of animals living in this zone.
- Write clue words about the conditions and include examples of organisms living in this zone: *underwater during high tide and exposed to air during low tide; clams and other shellfish, crabs, sea anemones, mussels.*
- Describe conditions in the zone. Choose one organism that lives here and write a short paragraph about it.

Lower Intertidal Zone:
- Draw a picture of animals living in this zone.
- Write clue words about the conditions and include examples of organisms living in this zone: *almost always covered with water; kelp and other seaweed thrive in strong waves; seaweed with root-like structures that anchor them to rocks; sea urchin, sea stars have suction-like feet.*
  Write clue words about tidal pools: *as waters recede, small pools of water left behind in cracks and openings; provide natural habitats for many plants and animals; shrimp burrow in sandy bottoms; most animals living here are invertebrates; adapted to living in water, open to air, and in harsh waves.*
- Describe conditions in the zone. Choose one organism that lives here and write a short paragraph about it.

Experiences, Investigations, and Research

Select one or more of the following activities for individual or group enrichment projects. Allow your students to determine the format in which they would like to report, share, or graphically present what they have discovered. This should be a creative investigation that utilizes your students' strengths.

1. Research and explain how animals have adapted to life in each of the intertidal zones.

2. Sketch the life cycle of animals found within each of the zones.

3. Determine what plants are found within the intertidal zone. Report on one.

Great Science Adventures

Lesson 14

What do we know about life on the seashore?

Ocean Concepts:
- The seashore is filled with animals and plants that depend on the ocean for existence; including seaweed, crabs, birds, and penguins.

Vocabulary: streamlined   algae   overlap   molt   *secrete

Construct and Read: *Lots of Science Library Book #14*

Activities:

Seashore – Graphic Organizer

Focus Skills: compare and contrast
Paper Handouts: *Discovering the Ocean* Graphic Organizer
a copy of Graphics 14A
Graphic Organizer: Glue Graphic 14A under the previous page of *Discovering the Ocean* at the glue line.

✎ Dictate or copy the name of each picture. Color the pictures.

✎✎ Title and write clue words for each picture: seaweed - *algae, cling to rocks with holdfasts, rubbery, travels with the tides.* hermit crab - *soft bodies and live in discarded shells,* sea gull - *hooked bill, webbed feet and long wings, eat shellfish and insects,* pelican - *use plunge-diving to catch fish, have pouch-like mouth to trap fish,* puffin - *live in large groups on cliffs, use wings to swim and catch fish,* penguin – *streamlined body, dark and white coloring, feathers overlap.*

✎✎✎ Describe life on the seashore. Explain the similarities and differences in the seashore animals.

Life in the Ocean – Graphic Organizer

Focus Skill: research
Paper Handouts: 12" x 18" piece of paper   8.5" x 11" piece of paper
a copy of Graphic 14B-C
Graphic Organizer: Fold the 12" x 18" piece of paper into a Shutter Fold. Title it *Life in the Ocean*. Cut out and glue chosen pictures from Graphic 14B on the cover. Use the 8.5" x 11" piece of paper to make a Large Question and Answer Book. Glue Graphic 14C on the cover and label it *Life on the Seashore*.

✎ Copy the name of a seashore animal on each tab. Under the tab, draw a picture of each animal.

✎✎ Choose two seashore marine animals to research. Title each tab with the name of your animal and under the tab write clue words about the animal.

✎✎✎ Choose two seashore marine animals to research. Title each tab with the name of your animal and under the tab write a paragraph about the animal.

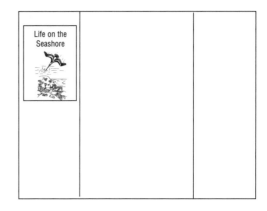

Glue the *Life on the Seashore* Large Question and Answer Book on the inside left tab of the *Life in the Ocean Graphic Organizer*, at the top.

Experiences, Investigations, and Research

Select one or more of the following activities for individual or group enrichment projects. Allow your students to determine the format in which they would like to report, share, or graphically present what they have discovered. This should be a creative investigation that utilizes your students' strengths.

1. Google: ocean seashore

2. Create a Time Line of major oil spills. Display the consequences of the oil spills on a map.

3. Investigate the importance of sand dunes on beaches.

4. Write a first person story of a sea turtle's life.

Great Science Adventures

Lesson 15

What are estuaries?

Ocean Concepts:
- Estuaries are partly enclosed bodies of water that contains fresh and ocean water.
- Coastal Plain Estuaries form when sea water rise and fill an exiting valley.
- Tectonic Estuaries form in folding and faulting land surfaces.
- Bar-built Estuaries when a lagoon or bay is flooded by ocean water
- Fjords are glacier valleys are flooded by seas or the ocean.

Vocabulary   estuary   lagoon   fjord   coastal   *tectonic

Construct and Read: Lots of Science Library Book #15

Activities:

Types of Estuaries

Focus Skills: comparing and contrasting
Paper Handouts: *Discovering the Ocean* Graphic Organizer
   a copy of Graphic 15A
Graphic Organizer: Glue Graphic 15A under the previous page of *Discovering the Ocean* at the glue line. Under each graphic:

✏️ Color the graphic and dictate or copy the name of each type of estuary.

✏️✏️ Complete ✏️. Write clue words about each type of estuary. Coastal Plain – *sea level rise and falls, fills up river valley.* Tectonic – *form by folding and faulting of land.* Bar-built – *shallow lagoon or bay flooded over a sand bar or barrier island.* Fjord – *glacier valleys folded by seas or ocean.*

✏️✏️✏️ Describe each type of estuary; how it is formed and examples of each. Include how the estuaries are alike and how the estuaries are different.

Make an Estuary

Materials: dirt   container of water
Activity: Use the dirt to build a type of estuary. Pour the water into it as it would happen from the ocean.

Experiences, Investigations, and Research

Select one or more of the following activities for individual or group enrichment projects. Allow your students to determine the format in which they would like to report, share, or graphically present what they have discovered. This should be a creative investigation that utilizes your students' strengths.

1. Google: estuary

2. Select and investigate an estuary. Explain it's past and present conditions.

3. Research the effects hurricanes have on estuaries.

4. List 3 reasons estuaries are important to life on Earth.

Great Science Adventures

Lesson 16

What do we know about crustaceans, mollusks, and sponges?

Ocean Concepts:
- Crustaceans are best known for their hard outer exoskeleton. They include lobsters, crab and shrimp.
- Mollusks are soft-bodied animals including snails, clams and sea slugs. A hard external shell is the most common characteristic of most mollusks.
- Mollusks can be univalves, one shell, or bivalves, two shells joined by a hinge.
- Other types of mollusks have no shell; these include octopus, squid, and cuttlefish.
- Sponges are brightly colored animals. Sponges are a network of openings and canals that connect to open pores. Water is pulled through this network to filter nutrients.

Vocabulary: crustacean   mollusk   sponge   canal   *univalves   *bivalves

Construct and Read: *Lots of Science Library Book #16*

Activities:

Crustaceans, Mollusks, and Sponges – Graphic Organizer

Focus Skills: explaining details and defining terms
Paper Handouts: *Discovering the Ocean* Graphic Organizer
 a copy of Graphics 16A
Graphic Organizer: Glue Graphic 16A under the previous page of *Discovering the Ocean* at the glue line.

✎ Color each picture and dictate or copy the name of each type of marine animal.

✎✎ Complete ✎. Write clue words for each picture: crustacean – *hard outer case, lobsters, shrimp, crabs, case must be replaced as animal grows, protect using claws and beaks,* mollusk – *soft-bodied, hard external shell, univalves have one shell, bivalves have two shells and a hinge sponge, some have no shell such as octopus and squid,* sponge – *brilliantly colored animal, network of canals for feeding.*

✎✎✎ Describe the sea animals on this graphic, explaining the details discovered in your reading. Define terms such as univalve, bivalve, and mantle.

Life in the Ocean – Graphic Organizer

Focus Skill: research
Paper Handouts: *Life in the Ocean Graphic Organizer*   8.5" x 11" piece of paper
   a copy of Graphic 16B
Graphic Organizer: Use the 8.5" x 11" piece of paper to make a Large Question and Answer Book. Glue Graphic 16B on the cover and label it *Crustaceans, Mollusks, and Sponges*.

✎   Copy the name of a crustacean, mollusk, or sponge on each tab. Under the tab, draw a picture of each animal.

✎✎   Choose two crustaceans, mollusks, or sponges to research. Title each tab with the name of your animal and under the tab write clue words about the animal.

✎✎✎   Choose two crustaceans, mollusks, or sponges to research. Title each tab with the name of your animal and under the tab write a paragraph about the animal.

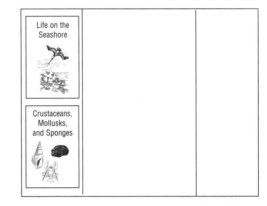

Glue the *Crustaceans, Mollusks, and Sponges* Large Question and Answer Book on the inside left tab of the *Life in the Ocean Graphic Organizer*, at the bottom.

Experiences, Investigations, and Research

Select one or more of the following activities for individual or group enrichment projects. Allow your students to determine the format in which they would like to report, share, or graphically present what they have discovered. This should be a creative investigation that utilizes your students' strengths.

1. Google: crustacean, mollusk, and ocean sponges.

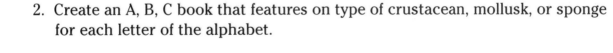
2. Create an A, B, C book that features on type of crustacean, mollusk, or sponge for each letter of the alphabet.

3. Investigate fossil mollusks.

Great Science Adventures

Lesson 17

What are the three types of fish?

Ocean Concepts:
- Bony fish make up about 96% of fish species.
- Bony fish have a skeleton of bone with a backbone and ribs. Most are covered with scales, have one pair of gills, and paired fins.
- Gills allow the fish to obtain oxygen from the water.
- Swim bladders in Bony fish allow them to be still and not drown.
- Cartilaginous fish have a skeleton make of cartilage and are covered with rough surface skin, not scales.
- Cartilaginous fish have no swim bladder so must move at all times or drown.
- Jawless fish are extinct except for hagfish and lampreys. Their bodies are long, tube-like, slimy and lack scales and paired fins.

Vocabulary: bony   cartilaginous   gills   swim bladder   scales   *pectoral   *dorsal   *anal

Construct and Read: *Lots of Science Library Book #17*

Activities:

Three Types of Fish – Graphic Organizer

Focus Skill: describing
Paper Handouts: *Discovering the Ocean* Graphic Organizer
a copy of Graphic 17A
Graphic Organizer: Glue Graphic 17A under the previous page of *Discovering the Ocean* at the glue line.

✎ Color each fish. Dictate or copy the names of the fins on the bony fish.
✎✎ Complete ✎. Write clue words for each type of fish: bony – *skeleton of bone, backbone, ribs, paired fins, swim bladder, scales, gills,* cartilaginous – *skeleton of cartilage, rough surface skin, paired fins, no swim bladder,* jawless – *extinct except for hagfish and lampreys, long, tube-like, slimy, no scales or paired fins.*
✎✎✎ Describe the three types of fish.

Life in the Ocean – Graphic Organizer

Focus Skill: research
Paper Handouts: *Life in the Ocean Graphic Organizer*   8.5" x 11" piece of paper
   a copy of Graphic 17B
Graphic Organizer: Use the 8.5" x 11" piece of paper to make a Large Question and Answer Book. Glue Graphic 17B on the cover and label it *Fish*.

✎ Copy the name of a fish on each tab. Under the tab, draw a picture of each fish.

✎✎ Choose two types of fish to research. Title each tab with the name of your animal and under the tab write clue words about the animal.

✎✎✎ Choose two types of fish to research. Title each tab with the name of your animal and under the tab write a paragraph about the animal.

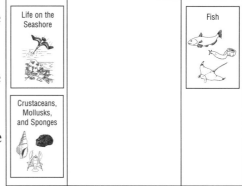

Glue the *Fish* Large Question and Answer Book on the inside right tab of the *Life in the Ocean Graphic Organizer*, at the top.

Experiences, Investigations, and Research

Select one or more of the following activities for individual or group enrichment projects. Allow your students to determine the format in which they would like to report, share, or graphically present what they have discovered. This should be a creative investigation that utilizes your students' strengths.

1. Google: bony fish, cartilaginous fish and jawless fish.

2. Investigate the whale shark and describe it as the world's largest fish.

3. Write an argument for or against the killing of sharks for sport.

Great Science Adventures

Lesson 18

What do we know about anemones, marine worms, echinoderms, and marine fishes?

Ocean Concepts:
- Sea anemones have symmetrical bodies, usually with stinging tentacles and a central mouth.
- Various types of marine worms are found throughout the ocean.
- Echinoderms are symmetrical creatures with a central mouth and include starfish, urchins, feather stars, and sea cucumbers.
- Marine fish are the most colorful sea animals, that live near coral reefs.

Vocabulary: anemone   symmetrical   tentacles   echinoderm   *paralyze   *entangle

Construct and Read: *Lots of Science Library Book #18*

Activities:

Anemones, Marine Worms, Echinoderms, and Marine Fishes – Graphic Organizer

Focus Skills: research and listing facts
Paper Handouts: *Discovering the Ocean* Graphic Organizer
a copy of Graphic 18A
Graphic Organizer: Glue Graphic 18A under the previous page of *Discovering the Ocean* at the glue line. Using the *Lots of Science Library Book #18:*

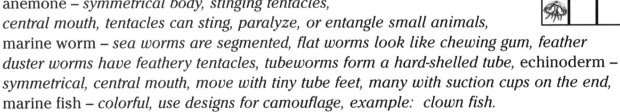

✏ Color and dictate or copy the name of each picture.
✏✏ Complete ✏. Write clue words for each picture:
anemone – *symmetrical body, stinging tentacles, central mouth, tentacles can sting, paralyze, or entangle small animals,*
marine worm – *sea worms are segmented, flat worms look like chewing gum, feather duster worms have feathery tentacles, tubeworms form a hard-shelled tube,* echinoderm – *symmetrical, central mouth, move with tiny tube feet, many with suction cups on the end,*
marine fish – *colorful, use designs for camouflage, example: clown fish.*
✏✏✏ Describe each type of animal on the graphics page. Research one type and write the information you find on the left side of the Graphic Organizer.

Life in the Ocean – Graphic Organizer

Focus Skill: research
Paper Handouts: *Life in the Ocean Graphic Organizer*   8.5" x 11" piece of paper
   a copy of Graphic 18B
Graphic Organizer: Use the 8.5" x 11" piece of paper to make a Large Question and Answer Book. Glue Graphic 18B on the cover and label it *Anemones, Marine Worms, Echinoderms, and Marine Fish.*

✎   Copy the name of an anemone, marine worm, echinoderm, or marine fish on each tab. Under the tab, draw a picture of each fish.

✎✎   Choose two types of anemones, marine worms, echinoderms, or marine fish to research. Title each tab with the name of your animal and under the tab write clue words about the animal.

✎✎✎   Choose two types of anemones, marine worms, echinoderms, or marine fish to research. Title each tab with the name of your animal and under the tab write a paragraph about the animal.

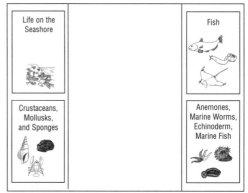

Glue the *Anemones, Marine Worms, Echinoderms, Marine Fish* Large Question and Answer Book on the inside right tab of the *Life in the Ocean Graphic Organizer*, at the bottom.

Experiences, Investigations, and Research

Select one or more of the following activities for individual or group enrichment projects. Allow your students to determine the format in which they would like to report, share, or graphically present what they have discovered. This should be a creative investigation that utilizes your students' strengths.

   1. Google: anemones, marine worms, and echinoderm.

Great Science Adventures

Lesson 19

What are coral reefs?

Ocean Concepts:
- A coral reef is a living colony of coral polyps.
- Coral polyps produce a hard outer skeleton to protect their soft bodies.
- Their cup-shaped bodies are ringed with stinging tentacles to help them catch food.
- Polyps can only live in warm waters.
- Most coral reefs grow around islands or along coastlines.
- A coral reef develops as polyps die and new polyps grow on top of their skeletons.
- Coral reefs protect the coastlines by breaking the waves, producing calm waters between the reef and beach.
- The three types of coral reefs are fringing reef, barrier reef, and atoll.

Vocabulary: coral reef   polyps   shapes   colors   skeleton   sunlight   warm   calm
*fringing reef   *barrier reef   *atoll

Construct and Read: *Lots of Science Library Book #19.*

Activities:

Coral Reefs - Graphic Organizer

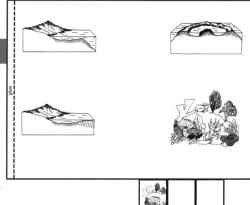

Focus Skill: comparing and contrasting
Paper Handouts: 8.5" x 11" paper   a copy of Graphics 19A
Graphic Organizer: Glue Graphic 19A under the previous page of *Discovering the Ocean* at the glue line. On the page:
✎   Draw a picture of each type of coral reef.
✎✎   Write clue words about each type of coral reef: Fringing reef - *grows out from coastline with no body of water separating land and reef*; Barrier reef - *lies farther from beach in deep waters; acts as a barrier, protecting land from ocean*; Atoll - *peaks of volcanoes surface above the water, creating an island; a coral reef can grow around the island; when a volcano sinks down below the surface of the water, a ring of coral called an atoll forms.*
✎✎✎   Complete ✎✎. Compare and contrast each type of coral reef.

Open all the tabs and on the bottom section:

✎ Draw a picture of a coral reef.

✎✎ Write clue words about how coral reefs form: *living colony of coral polyps; some coral look like plants; come in many shapes and colors; polyps continually divide and grow; polyps grow only a few inches a year; polyps produce hard outer skeleton to protect their soft bodies; cup-shaped bodies are ringed with stinging tentacles that help them catch food; most of their food is made by plant cells within polyps using sunlight, water, and carbon dioxide; can only live in warm waters that never fall less than 68 degrees F (20 degrees C); most coral reefs grow around islands or along coastlines; when polyps die, their hard skeletons remain and new polyps grow on top of them; coral reefs provide a habitat for many organisms; they protect coastlines by breaking the waves.*

✎✎✎ Describe coral reefs and explain how they form. Research the Great Barrier Reef and write a short paragraph.

Experiences, Investigations, and Research

Select one or more of the following activities for individual or group enrichment projects. Allow your students to determine the format in which they would like to report, share, or graphically present what they have discovered. This should be a creative investigation that utilizes your students' strengths.

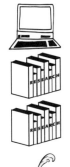

1. Google: coral reefs.

2. Locate and label the major coral reefs of the world.

3. Determine what water conditions are needed for a coral reef to be healthy and grow.

4. Google: Coral Reef Ecology Organizations. Compare the work of 2 or 3 organizations.

Great Science Adventures

Lesson 20

What do we know about sharks and rays?

Ocean Concepts:
- Sharks and rays have skeletons made of cartilage which allows them great flexibility in movement.
- Sharks have the most powerful jaws on Earth. Some eat fish, seals, and sea lions; others eat zooplankton.
- Baby sharks are called pups and can be born from eggs that hatch outside of the mother, eggs that hatch inside the mother, and pups that grow inside the mother.
- Rays are actually flattened sharks; some are capable of inflicting painful stings with their tails.

Vocabulary: shark   ray   pups   jaws   *flexibility   *denticle

Construct and Read: *Lots of Science Library Book #20*

Activities:

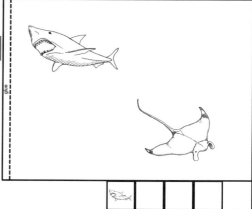

Sharks and Rays – Graphic Organizer

Focus Skills: describing details and processes
Paper Handouts: *Discovering the Ocean* Graphic Organizer
   a copy of Graphic 20A
Graphic Organizer: Glue Graphic 20A under the previous page of *Discovering the Ocean* at the glue line.

✏ Color and dictate or copy the name of each graphic.
✏✏ Complete ✏. Write clue words for each graphic:
shark – *most powerful jaws, shark bites with lower jaw and then upper jaw, tosses head back and forth to tear meat, teeth grow in parallel rows, cartilage skeleton, skin made of denticles, babies called pups,* ray – *flattened shark, moves gracefully through water, some have painful stings from their tails.*
✏✏✏ Describe the characteristics of sharks and rays; include how sharks bite prey and how pups are born.

Life in the Ocean – Graphic Organizer

Focus Skill: research
Paper Handouts: *Life in the Ocean Graphic Organizer*   3 pieces of 8.5" x 11" paper
   a copy of Graphic 20B-C
Graphic Organizer: Use the 3 pieces of 8.5" x 11" paper to make a 6 Tab Layered Look Book with tabs 1" deep. Cut off the bottom sheet to make the Layered Look Book 5 tabs. Glue Graphic 20B on the cover and label it *Life in the Ocean*. Glue 20C on the first tab. Under the cover:

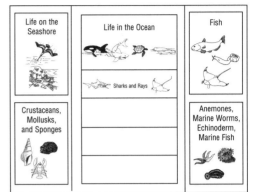

- ✎ Copy the name of a shark or ray on each tab and draw a picture for each fish.
- ✎✎ Choose two types of sharks or rays to research. Write clue words about the animals.
- ✎✎✎ Choose two types of sharks or rays to research. Write a paragraph about the animals.

Glue the Layered Look Book on the inside middle section of the *Life in the Ocean Graphic Organizer*.

Experiences, Investigations, and Research

Select one or more of the following activities for individual or group enrichment projects. Allow your students to determine the format in which they would like to report, share, or graphically present what they have discovered. This should be a creative investigation that utilizes your students' strengths.

1. Google: sharks and rays.

2. Compare sting rays and manta rays. How are they similar? How are they different?

3. Shark teeth are constantly replaced. Why is this important?

4. Research and report on legends based upon sharks and/or rays.

Great Science Adventures

Lesson 21

What do we know about marine reptiles?

Ocean Concepts:
- Marine reptiles include turtles and snakes. Both breathe air and must swim to the surface for air.
- Sea turtles mate at sea and migrate long distances to reach their breeding sites where 50 – 150 eggs are buried in a deep hole in the sand. About 2 months later, tiny turtles dig out of the hole.
- Sea turtles are the most endangered species in the world.
- Sea snakes are venomous but not aggressive.

Vocabulary: turtle   snake   blunt   *endangered   *venomous   *aggressive   *crevices

Construct and Read: *Lots of Science Library Book #21*

Activities:

Marine Reptiles – Graphic Organizer

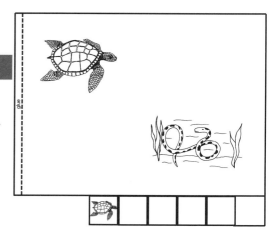

Focus Skill: locating information
Paper Handouts: *Discovering the Ocean* Graphic Organizer
a copy of Graphic 21A
Graphic Organizer: Glue Graphic 21A under the previous page of *Discovering the Ocean* at the glue line. Use *Lots of Science Library Book #21*:

✎   Color and dictate or copy a name for each picture.
✎✎   Complete ✎. Write clue words for each picture: sea turtle – *mates at sea, migrate to lay eggs in deep hole, 2 months later tiny turtles dig out of hole, only 1 in 1,000 will live to be adult, eat jellyfish, sponges, crabs, squids, and fishes,* sea snake – *venomous, not aggressive, small blunt head, scales, and flattened head, eat fishes, give birth to fully developed young*
✎✎✎   List facts about each marine reptile. Use research materials to find facts that are not included in *Lots of Science Library Book #21*

Life in the Ocean - Graphic Organizer

Focus Skill: research
Paper Handouts: *Life in the Ocean Graphic Organizer*
    a copy of Graphic 21B
Graphic Organizer: Glue Graphic 21B on the bottom of the second tab. Open that tab and:

- ✏ Copy the name of a marine reptile and draw a picture of a reptile.
- ✏✏ Choose two types of marine reptiles to research. Write clue words about the animals.
- ✏✏✏ Choose two types of marine reptiles to research. Write a paragraph about the animals.

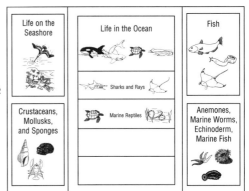

Experiences, Investigations, and Research

Select one or more of the following activities for individual or group enrichment projects. Allow your students to determine the format in which they would like to report, share, or graphically present what they have discovered. This should be a creative investigation that utilizes your students' strengths.

1. Google: marine reptile

2. List and describe organizations that protect sea turtles, their habitats, and nesting areas.

3. Compare and contrast sea snakes and venomous land snakes.

4. Sketch to scale several species of sea turtles. Try to determine if there is a relationship between size and level of endangerment.

Great Science Adventures

Lesson 22

What do we know about marine mammals?

Ocean Concepts:
- Marine mammals are warm-blooded, give birth to living young, and breathe air through lungs.
- Marine mammals include whales, dolphins, sea otters, and seals.
- Marine mammals show remarkable abilities to communicate and learn.

Vocabulary: mammal   warm-blooded   dolphin   orca   *sonar   *baleen

Construct and Read: *Lots of Science Library Book #22*

Activities:

Marine Mammals – Graphic Organizer

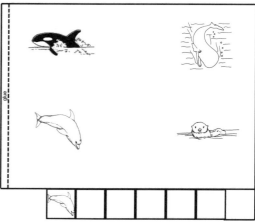

Focus Skill: explaining characteristics
Paper Handouts: *Discovering the Ocean* Graphic Organizer
   a copy of Graphics 22A
Graphic Organizer: Glue Graphic 22A under the previous page of *Discovering the Ocean* at the glue line.

✎   Color and dictate or copy a name for each picture.
✎✎ Complete ✎. At the top, list the characteristics of mammals: *warm-blooded, give birth to living young, breathe through lungs.* Write clue words for each picture: blue whale – *Earth's largest animal, eats krill, a zooplankton, baleen filters food from water, breathes through blowhole at top of its head* sperm whale – *the largest toothed whale, large head, eats squid, fish and sharks* orca – *known as Killer Whale, travels in pods, eats whales, dolphins, seals, and fish,* dolphin – *can dive 1,000 feet below surface, jumps up to 20 feet out of water, hunts for fish near surface, lives in pods,* sea otter – *hunts on ocean floor, returns to surface to eat while floating on back.*
✎✎✎ Explain the characteristics and living patterns of each animal.

Life in the Ocean – Graphic Organizer

Focus Skill: research
Paper Handouts: *Life in the Ocean Graphic Organizer*
a copy of Graphic 22B
Graphic Organizer: Glue Graphic 22B on the bottom of the third tab. Open the tab and:

 Copy the name of a marine mammal and draw a picture of it.

 Choose two types of marine mammals to research. Write clue words about the animals.

 Choose two types of marine mammals to research. Write a paragraph about the animals.

Experiences, Investigations, and Research

Select one or more of the following activities for individual or group enrichment projects. Allow your students to determine the format in which they would like to report, share, or graphically present what they have discovered. This should be a creative investigation that utilizes your students' strengths.

1. Google: marine mammals.

2. Create and illustrate a time line that outlines the importance of whales to humankind.

3. List and describe products once obtained from whales.

4. What is ocean sound pollution? Where does it come from and how might it effect marine mammals?

Great Science Adventures

Lesson 23

What do we know about pinnipeds?

Ocean Concepts:
- Pinnipeds are mammals with front and hind flippers.
- Types of pinnipeds include seals, sea lions, and walruses.

Vocabulary: seal   sea lion   walrus   pinniped   flippers   tusks

Construct and Read: *Lots of Science Library Book #23*

Activities:

Pinnipeds – Graphic Organizer

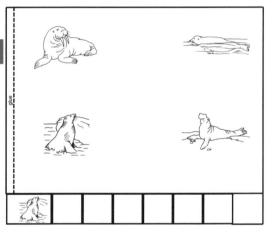

Focus Skill: listing information
Paper Handouts: Discovering the Ocean Graphic Organizer
   a copy of Graphics 23A
Graphic Organizer: Glue Graphic 23A under the previous page of *Discovering the Ocean* at the glue line.

✏ Color and dictate or copy a name for each picture.
✏✏ Complete ✏. At the top, list the characteristics of pinnipeds: *mammals with front and back pairs of flippers*. Write clue words about each picture: seal – *sleeps with body submerged in water and tip of nose out to breath, eats fish, squid, and crustaceans*, sea lion – *found on surface with head and flippers out of water to absorb heat, makes noisy barks*, elephant seal – *second best diver in ocean, can dive for about 30 minutes, replaces old skin and hair at one time, migrates to Northern Pacific for feeding*, walrus – *large size, big tusks, whiskers, tusks can be 31 – 39 inches long, eats clams, worms, crustaceans, fish and young seals*.
✏✏✏ List information about each type of pinniped.

Life in the Ocean – Graphic Organizer

Focus Skill: research
Paper Handouts: *Life in the Ocean Graphic Organizer*
   a copy of Graphic 23B
Graphic Organizer: Glue Graphic 23B on the bottom the fourth tab of the Layered Look Book. Open the tab:

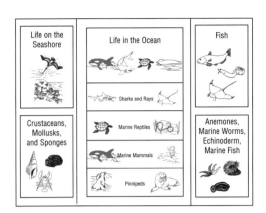

- Copy the name of a pinniped and draw a picture.
- Choose two types of pinnipeds to research. Write clue words about the animals.
- Choose two types of pinnipeds to research. Write a paragraph about the animals.

Experiences, Investigations, and Research

Select one or more of the following activities for individual or group enrichment projects. Allow your students to determine the format in which they would like to report, share, or graphically present what they have discovered. This should be a creative investigation that utilizes your students' strengths.

1. Google: pinnipeds.

2. What countries allow seal hunting? Why are seals hunted? Why do some countries ban products made from pennipeds?

3. Compare walrus and elephant tusks.

4. Which pennipeds live in cold regions and warm regions? Record this information on a map.

Great Science Adventures
Lesson 24

What are ocean resources?

Ocean Concepts:
- About one-fifth of the world's oil and natural gas comes from beneath the ocean floor.
- Currently, many minerals in the ocean are difficult and expensive to extract.
- Modern technology has caused overfishing in many areas.
- Oil Spills cause damage in the ocean and on the seashore.

Vocabulary: oil   natural gas   drill

Construct and Read: *Lots of Science Library Book #24.*

Activities:

Ocean Resources - Graphic Organizer

Focus Skill: describing
Paper Handouts: a piece of 8.5" x 11" paper
a copy of Graphics 24A-B
Graphic Organizer: Make a Large Question and Answer Book. Title the cover *The Ocean Today*. Glue Graphic 24A on the first tab, title it *Ocean Resources*. Glue Graphic 24B on the second tab, title it *Problems in the Ocean*. Under each tab:
✏ Draw the picture from the tab.
✏✏ Write clue words: resources - *oil, natural gas, salt, tin, seafood, fishing and swimming*
    problems - *oil spillage, modern technology causes overfishing, pollution*
✏✏✏ Explain the resources we use from the ocean. Describe the problems that have been created over the years of use. Research methods of solving two of these problems.

Glue *The Ocean Today* Large Question and Answer Book on the back of the *Life in the Ocean* Desktop Project.

Experiences, Investigations, and Research
Select one or more of the following activities for individual or group enrichment projects. Allow your students to determine the format in which they would like to report, share, or graphically present what they have discovered. This should be a creative investigation that utilizes your students' strengths.

1. Google: ocean resources, overfishing, and oil spills.

Notes

Notes

Great Science Adventures

Lots of Science Library Books

Each *Lots of Science Library Book* is made up of 16 inside pages, plus a front and back cover. All the covers to the *Lots of Science Library Books* are located at the front of this section. The covers are followed by the inside pages of the books.

How to Photocopy the *Lots of Science Library Books*

As part of their *Great Science Adventure*, your students will create *Lots of Science Library Books*. The *Lots of Science Library Books* are provided as consumable pages which may be cut out of the *Great Science Adventures* book at the line on the top of each page. If, however, you wish to make photocopies for your students, you can do so by following the instructions below.

To photocopy the inside pages of the *Lots Of Science Library Books*:

1. Note that there is a "Star" above the line at the top of each *LSLB* sheet.

2. Locate the *LSLB* sheet that has a Star on it above page 16. Position this sheet on the glass of your photocopier so the side of the sheet which contains page 16 is facing down, and the Star above page 16 is in the left corner closest to you. Photocopy the page.

3. Turn the *LSLB* sheet over so that the side of the *LSLB* sheet containing page 6 is now face down. Position the sheet so the Star above page 6 is again in the left corner closest to you.

4. Insert the previously photocopied paper into the copier again, inserting it face down, with the Star at the end of the sheet that enters the copier last. Photocopy the page.

5. Repeat steps 1 through 4, above, for each *LSLB* sheet.

To photocopy the covers of the *Lots of Science Library Books*:

1. Insert "Cover Sheet A" in the photocopier with a Star positioned in the left corner closest to you, facing down. Photocopy the page.

2. Turn "Cover Sheet A" over so that the side you just photocopied is now facing you. Position the sheet so the Star is again in the left corner closest to you, facing down.

3. Insert the previously photocopied paper into the copier again, inserting it face down, with the Star entering the copier last. Photocopy the page.

4. Repeat steps 1 through 3, above, for "Cover Sheets" B, C, D, E, and F.

Note: The owner of this book has permission to photocopy the *Lots of Science Library Book* pages and covers for classroom use only.

How to assemble the *Lots of Science Library Books*

Once you have made the photocopies or cut the consumable pages out of this book, you are ready to assemble your *Lots of Science Library Books*. To do so, follow these instructions:

1. Cut each sheet, both covers and inside pages, on the solid lines.

2. Lay the inside pages on top of one another in this order: pages 2 and 15, pages 4 and 13, pages 6 and 11, pages 8 and 9.

3. Fold the stacked pages on the dotted line, with pages 8 and 9 facing each other.

4. Turn the pages over so that pages 1 and 16 are on top.

5. Place the appropriate cover pages on top of the inside pages, with the front cover facing up.

6. Staple on the dotted line in two places.

You now have completed *Lots of Science Library Books*.

What is oceanography?

Lots of Science Library Book #4

What do we know about the ocean?

Lots of Science Library Book #3

What do we know about water?

Lots of Science Library Book #2

What do we know about Earth?

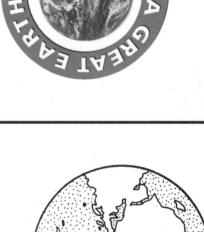

Lots of Science Library Book #1

★

oceanography study plants animals descend *sonar *submersibles	water solid liquid gas density water cycle *hydrosphere *hydrogen *oxygen *surface tension *evaporation *condensation *precipitation
oceans Pacific Ocean Atlantic Ocean Indian Ocean Arctic Ocean Southern Ocean seas rain salty *deltas	Earth Sun solar system continents *continental plates *Pangaea

What lives in the ocean layers?

Lots of Science Library Book #8

What are the ocean layers?

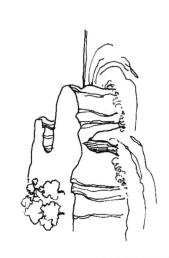

Lots of Science Library Book #7

What do we know about the ocean floor?

Lots of Science Library Book #6

What is saltwater?

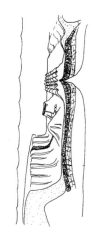

Lots of Science Library Book #5

B

surface dwellers
free swimmer
bottom dweller

*phytoplankton

---

temperature
depth
radiates

*Sunlite Zone
*Twilight Zone
*Dark Zone
*Abyss
*Trenches

---

slope
downward
shallow
valleys
trenches
mountains
steep

*continental shelf
*continental slope
*abyssal plains
*abyssal hills
*oceanic ridge

---

salt
minerals
float
salinity
microorganisms

*halite
*sodium chloride
*calcium sulfate
*calcium carbonate
*desalinization

How does the ocean affect weather?

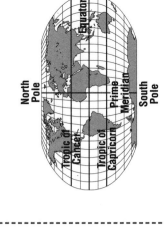

Lots of Science Library Book #12

What are ocean currents?

Lots of Science Library Book #11

What are tides?

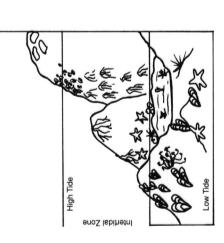

High Tide
Low Tide
Intertidal Zone

Lots of Science Library Book #10

What are waves?

Lots of Science Library Book #9

Top-right quadrant:
- weather
- hot
- cold
- hurricane
- tsunamis
- storm surges

Top-left quadrant:
- currents
- surface
- rotation
- winds
- land masses
- saltiness

Bottom-right quadrant:
- high
- far
- gravity
- daily
- time
- size
- high tide
- low tide
- Sun
- moon
- *spring tide
- *neap tide

Bottom-left quadrant:
- moving
- waves
- winds
- speed
- forward
- shape
- *fetch
- *transverse waves
- *trough
- *crest
- *breaker
- *tsunami

What do we know about crustaceans, mollusks, and sponges?

Lots of Science Library Book #16

What are estuaries?

Lots of Science Library Book #15

What do we know about life on the seashore?

Lots of Science Library Book #14

What is the intertidal zone and what lives there?

Lots of Science Library Book #13

Top-right quadrant:

crustacean
mollusk
sponge
canal
*univalve
*bivalve

Top-left quadrant:

estuary
lagoon
fjord
coastal
*tectonic

Bottom-right quadrant:

streamlined
algea
overlap
molt
*secrete

Bottom-left quadrant:

lower
middle
upper
splash
predator
prey
air
exposure
water
*ebb tide
*intertidal zone
*tidal pools

What do we know about sharks and rays?

Lots of Science Library Book #20

What are coral reefs?

Lots of Science Library Book #19

What do we know about anemones, marine worms, echinoderms, and marine fishes?

Lots of Science Library Book #18

What are the three types of fish?

Lots of Science Library Book #17

shark
rays
pups
jaws

*flexibility
*denticle

anemone
symmetrical
tentacles
echinoderm

*paralyze
*entangle

coral reef
polyps
shapes
colors
skeleton
sunlight
warm
calm

*fringing reef
*barrier reef
*atoll

bony
cartilaginous
gill
swim bladder
scales

*pectoral
*dorsel
*anal

What are ocean resources?

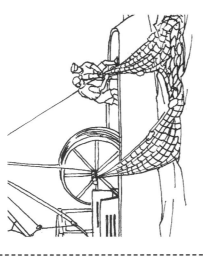

Lots of Science Library Book #24

What do we know about pinnipeds?

Lots of Science Library Book #23

What do we know about marine mammals?

Lots of Science Library Book #22

What do we know about marine reptiles?

Lots of Science Library Book #21

oil
natural gas
drill

seal
sea lion
walrus
pinniped
flippers
tusks

mammal
warm-blooded
dolphin
orca
*sonar
*baleen

turtle
snake
blunt
*aggressive
*endangered
*venomous
*crevices

5

12

Most scientists believe that at one time all the continents were joined together, forming one massive land formation called Pangaea.

7

10

The land masses that separate the oceans are called continents: North America, South America, Europe, Asia, Africa, Oceania, and Antarctica. Look at a world map. The continents look as though they could fit together like a puzzle.

1

We live on planet Earth, the third planet from the Sun. Earth, eight other planets, and the Sun make up our solar system.

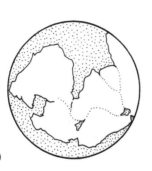

The solar system consists of the planets Mercury, Venus, Earth, Mars, Jupiter, Saturn, Uranus, Neptune, and Pluto; their moons; and other celestial bodies. Earth is about 93 million miles (150 million km) from the Sun.

3

16

Scientists believe that the continental plates shifted over time to position the continents where they are today.

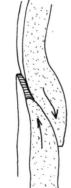

14

If two continental plates move apart, molten rock comes up, cools, and forms new rock. This can happen on land or on the ocean floor. Molten rock is called magma.

An axis is an imaginary straight line on which a planet rotates.

Some planets contain very small amounts of water in the form of ice.

Earth is surrounded by an atmosphere, a layer of gases which also contains water vapor. Although the amount of water vapor in Earth's atmosphere is very small, without it we would not have clouds, rain, or snow. The Earth's atmosphere consists of less than 1% water vapor.

Note: Some scientists recognize only the Pacific, Atlantic, Indian, and Arctic Oceans.

From space, water vapor makes the Earth's surface appear blue with swirls of white clouds. Most of Earth's water is found in the oceans: Pacific Ocean, Atlantic Ocean, Indian Ocean, Arctic Ocean, and Southern (or Antarctic) Ocean.

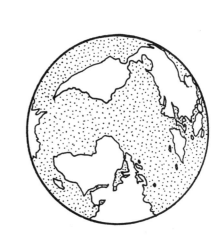

The Sun is at the center of our solar system. It produces light and heat from its core. All the planets revolve around, or circle, the Sun and rotate, or spin, on their own axes.

The Earth's crust is divided into large pieces called continental plates. Continental plates move very slowly, like rafts on water. The movement of these plates is called plate tectonics.

If two continental plates collide, they will either both be pushed upward or one may be pushed underneath the other. Continental plates can also move sideways against one another.

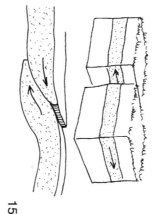

Earth is different from any other planet in the solar system because its surface contains water in liquid form. About 75% of the Earth's surface is covered with water.

Have you ever noticed how ice floats in a glass of water? Ice floats because water is less dense in its solid state. Density is the amount of matter in a substance compared to its volume.

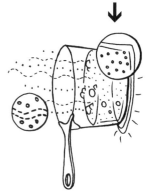
iceberg

5

Water molecules in liquid form do not remain locked together as a solid, but they still move slowly enough to be attached to each other.

7

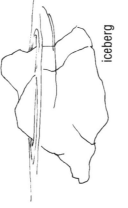

12

Heat from the Sun turns water in the oceans, lakes, rivers, and ground into water vapor. Water vapor rises into the atmosphere through evaporation. As water vapor meets cooler air, condensation occurs, forming clouds. When the clouds cannot hold any more water vapor, it turns back into water droplets that fall as precipitation.

Lots of Science Library Book #2   10

All the water on and near the Earth's surface is called the hydrosphere. The word hydrosphere comes from the Greek words *hydro*, meaning "water" and *sphere*, meaning "round."

Lots of Science Library Book #2   1

Water occurs in three states: solid, liquid, and gas. It is the only substance on Earth that occurs naturally in all three states.

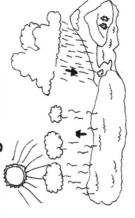

3

★

Wonderful Wonders in the Water

Sea horses are between 2-6 inches long. Sea horses have fins and gills. The male sea horse has a pouch on the underside of his body. The female sea horse places her eggs in this pouch where the male protects them until they hatch.

Lots of Science Library Book #2   16

Groundwater makes up about .5% of the Earth's water. Rivers, lakes, and streams make up about .02%. The water vapor in the atmosphere consists of about .0001% of the Earth's water.

Lots of Science Library Book #2   14

Water molecules in water vapor, a gas, are so far apart and move so quickly that they rarely collide and do not become attached to one another.

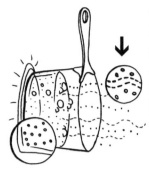

Another unique quality of water is the molecules' ability to stick to each other most strongly at the surface, producing a "skin" called surface tension. This is why water, when placed on a flat surface, looks like a dome.

Water molecules' ability to stick together is called cohesion. Water's ability to stick to other molecules is called adhesion.

Water, in all three forms, is constantly moving over the surface of the Earth. Water continually changes from one form to another by evaporation, condensation, and precipitation. This is called the water cycle.

The water cycle is also called the hydrologic cycle.

Water is made up of hydrogen (H) and oxygen (O). It is scientifically written as H_2O, which indicates that a water molecule consists of two hydrogen atoms and one oxygen atom.

Water molecules in ice, a solid, are locked together in hexagonal, or six-sided, crystals. Water molecules vibrate quickly, but not fast enough to break out of the crystal.

More than 97% of the Earth's water is found in the oceans. Ice on Antarctica and glaciers make up about 2% of the Earth's water.

Oceans

Underground rivers, lakes, & streams

Antarctica and glaciers

The Pacific Ocean, the world's largest and deepest ocean, is larger than all the Earth's land combined and holds about 50% of all the Earth's water.

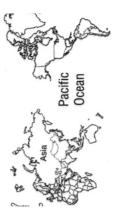

The Atlantic Ocean is the second-largest ocean, covering 40 million sq. mi. (104 million sq. km). It reaches from North and South America to Europe and Africa. The deepest parts of the Atlantic Ocean are recorded at about 30,000 ft (9.15 km), with an average depth of about 11,000 ft. (3.35 km).

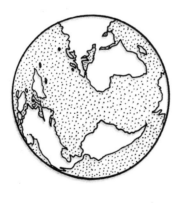

The Southern or Antarctic Ocean is made up of very cold water that flows around Antarctica in a clockwise, or westerly, direction. Currents move 185 million tons of ocean water every second. This ocean covers 11 million sq. mi. (28 million sq. km)

More than 97% of the Earth's water is found in the ocean.

Arctic Ocean
Atlantic Ocean
Pacific Ocean
Indian Ocean
Southern Ocean

Rain and snow fall onto land and eventually flow downhill into rivers and finally into the sea. As rivers flow into seas, they often drop soil and other materials, forming triangular-shaped banks called deltas.

Delta comes from the Greek letter delta Δ

The energy from the Sun keeps Earth's water in constant motion. Water in the ocean evaporates and eventually returns to Earth in the form of rain or snow.

It stretches from North Asia and South America to Asia and Australia, covering about 69 million sq. mi. (179 million sq. km). Areas of the Pacific Ocean plunge to about 36,000 ft. (11 km), with an average depth of about 14,000 ft. (4.27 km).

The Indian Ocean covers about 28 million sq. mi. (72 million sq. km). The average depth is about 13,000 ft (3.97 km), and the deepest areas are about 25,000 ft. (7.62 km).

The Arctic, the smallest, shallowest, and coldest ocean, mostly covered with ice, is located at the North Pole and stretches about 5,440,000 sq. mi. (14 million sq. km). The average depth is about 4,265 ft. (1.3 km), with a maximum depth of about 17,880 ft. (5.45km).

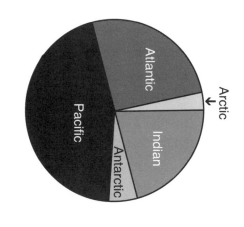

Most scientists theorize that Pangaea shifted, eventually creating seven continents. These continents divided the Earth's water into five oceans we know today: Pacific, Atlantic, Indian, Arctic, and Southern (or Antarctic) Oceans.

Oceans contain about 97% of Earth's water, more than 1 quintillion (1,500,000,000,000,000,000) tons of water. The average depth of the oceans is about 12,447 ft. (3794 m).

Rainwater continually flows into rivers, picking up salts and other minerals along the way, and eventually dumps them into the ocean. This cycle keeps the water in the ocean salty.

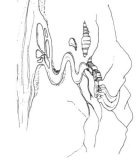

Seas are smaller bodies of salt water that are usually partly or completely surrounded by land. Seas of the world include Caribbean Sea, Mediterranean Sea, North Sea, and Red Sea. The largest sea in the world is the South China Sea in the Pacific Ocean.

1

The early explorations of men such as Columbus, da Gama, Magellan, and Cook provided vital information to map the world.

3

The expedition was headed by a Scotsman, C. W. Thomson, and a Canadian geologist, John Murray. The term oceanography, the study of the oceans and seas, originated from their expedition.

5

Oceanography covers five main areas:

1) the study of the ocean floor;

2) the physical characteristics of ocean water, such as currents and temperature;

7

Today, there are several institutions of oceanography around the world. Modern instruments such as sonar, diving equipment, and submersibles have greatly increased our ability to observe and understand the oceans.

10

The depth of the ocean at a given location can be determined by measuring how long it takes for sound waves to reach the ocean floor and come back again.

12

Diving, or scuba, equipment gives divers freedom of movement while exploring underwater. Scuba divers can safely descend to only about 165 ft (50 m).

14

A submersible is a small craft housed in a larger submarine. A submersible can descend to deeper waters than a submarine.

Since 1964, ALVIN, a three-person submersible, has made thousands of successful trips, reaching depths of up to 13,120 ft (4,000 m). Pressure at these depths would crush ordinary submarines.

16

Wonderful Wonders in the Water

As a tiny larva, the mollusk builds its shell by depositing calcium. As the creature grows, the shell is extended outward to create a perfect spiral.

3) the chemical properties of ocean water, such as saltiness, or salinity;

4) the study of oceanic plants and animals;

5) the ocean's interaction with weather.

Oceanographic institutions in the United States include Woods Hole in Massachusetts and Scripps Institution in California.

Sonar is a device that uses sound waves to detect the location and distance of objects. Sonar is used to measure depths of water and to map the ocean floor.

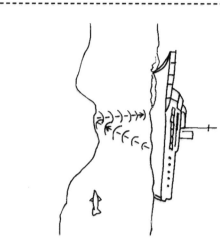

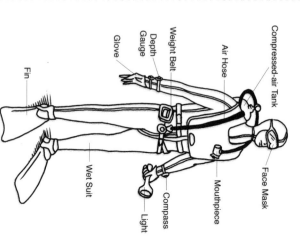

As navigation continued and navigational tools improved, people gained a better understanding of the oceans. However, it was not until 1872, aboard the H.M.S. Challenger, that official oceanic studies began.

The Challenger's mission was to study the ocean by observing and recording water temperatures, current movements, plants and animals. In the course of the study scientists collected rocks and mud samples, and performed many other procedures.

Page 5

Page 7

The salt content of the Dead Sea is so high it cannot sustain life except for a few types of microorganisms. This is why it was named the Dead Sea.

Page 12

Since salt is left behind during evaporation, rain and snow fall as fresh water.

Page 10

When evaporation takes place, the salt remains behind. The climate in the Dead Sea region is hot, causing the sea water to evaporate rapidly. This creates a higher salinity.

Page 1

Have you ever been in the ocean and tasted the water? The water that makes up Earth's ocean contains salts, including the mineral halite.

Page 3

Page 16

A quart of ocean water may contain millions of bacterial cells, hundreds of thousands of phytoplankton, and tens of thousands of zooplankton.

Page 14

Desalinization involves heating saltwater until it evaporates, leaving the salt behind. In this process, water vapor moves over cold pipes. As the hot water vapor comes into contact with the cold air around the pipes, condensation occurs, and salt-free water is formed.

Some bodies of water contain larger amounts of salt than others. The degree of saltiness of water is called its salinity. The Dead Sea, which is actually a saltwater lake, is about seven times as salty as the ocean.

Salinity decreases where rivers meet oceans, because fresh water from the rivers dilutes the salty sea water. Very cold bodies of water, such as the Arctic Ocean, also have lower salinity than warmer waters. This is due to slower evaporation and the addition of fresh water from melting ice.

Ocean water has a mineral composition of sodium chloride, that is the same substance that is in table salt. An average of six heaping tablespoons of salt are in a gallon of sea water.

Sodium chloride is NaCl. The average amount of salt in sea water is 35 parts per thousand, or 35 ppt.

Density is the amount of matter in a substance compared to its volume. The greater the salinity of the water, the greater the density of the water. The density of sea water depends on its salinity and temperature. If you have been in the ocean, you probably noticed that it is easier to float in saltwater than in freshwater.

The more dense the water, the easier it is to float in it. The water in the Dead Sea is so dense that it is almost impossible to sink in it.

Other salts, such as calcium sulfate and calcium carbonate, dissolve or settle on the ocean floor. Many sea animals use these minerals to produce shells and bones.

Because ocean water is salty, it cannot be used for drinking. However a process called desalinization can be used to separate the salt from the saltwater.

This process is used in some countries that do not have enough fresh water, such as Saudi Arabia and Kuwait. Unfortunately, desalinization is a costly process, and poorer countries are unable to afford it.

The majority of the world's seafood is harvested in the Sunlit Zone of the oceans.

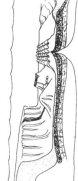

5

After the shelf break, the ocean ground takes a steep downward slope to the depths of the ocean floor.

7

Valleys, called trenches, are also found on the ocean floor. The deepest part of Earth's surface is the Marianas Trench in the northwest Pacific Ocean. If a large stone were dropped from the ocean surface in the Marianas Trench, it would take about one hour to reach the bottom of the trench.

12 Lots of Science Library Book #6

Beyond the abyssal plains are ridges of abyssal hills which make up about one-third of the ocean floor. Some of these reach above the water, creating islands. Islands such as Iceland, the Azores, and the Galapagos Islands are the tops of oceanic mountain ranges.

10 Lots of Science Library Book #6

Walking on the beach and into the ocean is similar to walking from the shallow end of a pool to the deeper end. When you move from land to the ocean, the ground begins to slope downward and the water becomes deeper.

Lots of Science Library Book #6 1

The continental shelf gradually slopes downward to about 650 ft. (200 m). The waters above the shelf, called the Sunlit Zone, make up about 10% of the ocean's water.

Sunlit Zone

3

Hydrothermal vents are usually about 33 feet (10 m) across. Their sizes range from the size of a pool table to the size of a tennis court.

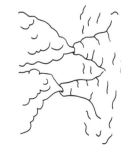

16 Lots of Science Library Book #6

In some areas of the ocean floor, plates of Earth's crust move away from each other allowing magma to well up in the cracks. The magma cools and solidifies adding to the ocean floor. According to scientists, the Atlantic Ocean floor is widening about 2 inches (5 cm) each year.

14 Lots of Science Library Book #6

The average continental shelf is 430 ft. (130 m) into the ocean and consists of huge deposits of sand, mud, and gravel. The edge of the continental shelf is called the shelf break.

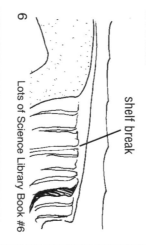
shelf break

Continental slopes end in smooth surfaces called the abyssal plains. These plains are the flattest places on Earth and cover approximately one-half of the deep ocean floor. The water above the abyssal plains is about 6,500 – 13,000 feet deep (2,000 – 4,000 m), black and very cold.

shelf
slope

In the Pacific Ocean there are over 14,000 mountains that are 2,000 – 6,000 ft (210 – 1,829 m) high yet remain completely below the ocean surface. Using sonar, sound waves which record distance, scientists measure the depths of the ocean.

The slope on the ocean floor is called the continental shelf. This shelf surrounds all the continents forming a type of ledge.

continental shelf

Sun filters through the waters above the continental shelf making it well suited for plants to grow in abundance. These plants provide food and shelter for animals of various sizes.

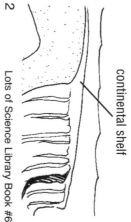

Scientists on ALVIN recently discovered hydrothermal vents, or black smokers. Black smoke and sulfur-rich hot water blast out of the tops of these chimney-like structures.

Parts of the Marianas Trench are about 35,000 ft (11,000 m) deep, six times as deep as the Grand Canyon. The Marianas Trench stretches about 1,550 miles (2,500 km) long.

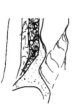

Wonderful Wonders in the Water

When the tide of southern Africa is out, the plough snail burrows into the sand. When the tide comes in, it comes to the surface sucking water into its foot. The water sweeps the snail high up on the shore to find food.

The ocean includes areas of various temperatures and depths. The Sun radiates heat and light to Earth, warming the surface of the ocean.

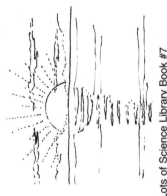

Beyond 19,800 feet (6,000 m) lie lower areas in trenches and canyons. Even with the cold temperatures and blackness of the water, life can be found.

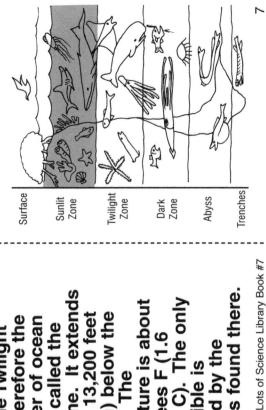

Surface
Sunlit Zone
Twilight Zone
Dark Zone
Abyss
Trenches

The next layer, named the Abyss, extends from 13,200 feet (4,000 m) to 19,800 feet (6,000 m) below the ocean's surface. The water is near freezing and there is no light at all. The mud can be more than a mile thick and is made from the skeletons of small sea animals.

Sunlight cannot filter through the ocean below the Twilight Zone; therefore the next layer of ocean water is called the Dark Zone. It extends down to 13,200 feet (4,025 m) below the surface. The temperature is about 35 degrees F (1.6 degrees C). The only light visible is produced by the creatures found there.

The Sunlit Zone consists of the top 600 feet (183 m). The water is warm due to the heat of the Sun, and most of the plants and animals live in this area of the ocean.

The Twilight Zone is the next 2,700 feet (823 m). The temperature can be as low as 41 degrees F (5 degrees C) because of the lack of sunlight.

Surface waters near the North and South Poles reach only about 29 degrees F (1.6 degrees C). Near the equator the surface waters can reach up to 85 degrees F (29.4 degrees C). However, even near the equator, only the top few hundred feet of water is warmed.

Ocean waters are divided into five layers:

1. Sunlit Zone
2. Twilight Zone
3. Dark Zone
4. Abyss
5. Trenches

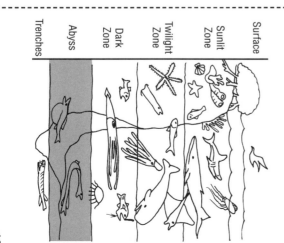

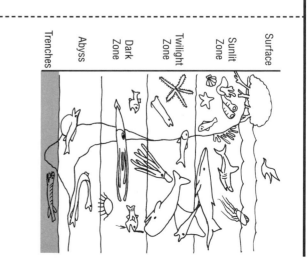

Phytoplankton makes up about 99% of plant life in the ocean.

Red tide occurs when phytoplankton becomes so numerous the water looks reddish. The abundance of phytoplankton depletes oxygen from the water.

5

Surface drifters are unable to propel themselves effectively and are therefore moved about by the wind or current. They live in the Sunlit Zone of the ocean.

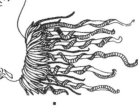

7

In the Twilight Zone, with little if any sunlight filtering into this area, plants cannot grow and creatures may create their own light.

12   Lots of Science Library Book #8

Free swimmers make up the largest group of marine animals in the ocean. Free swimmers are able to propel themselves through the water to find food and defend themselves. This group includes fish, mammals, and reptiles. Most free swimmers live in the Sunlit Zone where food is abundant.

10   Lots of Science Library Book #8

Oceans are teeming with plant and animal life of all sizes.

1   Lots of Science Library Book #8

Rooted plants, such as kelp and seaweed, are only capable of growing in the Sunlit Zone where there is enough light for photosynthesis to take place. Plants of this type make up 1% of the ocean's plants.

3

The Abyss and trenches are nearly freezing and totally dark. Free swimmers, such as tube worms, and bottom dwellers, such as starfish, are able to survive at these levels.

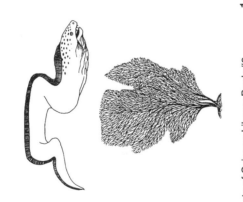

16   Lots of Science Library Book #8

A female anglerfish has a lure extending from its head that contains light-producing bacteria. The glowing light attracts an easy meal for the anglerfish.

14

Marine animals are divided into three groups: surface drifters, free swimmers, and bottom dwellers.

Zooplankton are drifting animals that include microscopic protozoa and small crustaceans. They float on or near the surface of the water, drifting with the currents. Although zooplankton usually very small, zooplankton can grow to a large size.

Jellyfish and Portuguese man-of-war are larger types of zooplankton that are still unable to propel themselves.

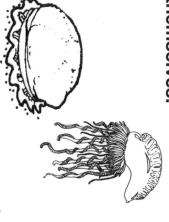

Two types of plants are found in the ocean, those with roots attached to the ocean bottom and those that drift in the water.

The most abundant plants in the ocean are single-celled, minute floating plants called phytoplankton. In a bucket of sea water, there can be millions of these microscopic relatives of seaweed. Phytoplankton use sunlight and nutrients from the sea to grow.

Light producing bacteria help fish attract mates, find prey, and defend themselves. Hatchet fish have lighted organs on the underside of their bodies, confusing predators.

Light producing free swimmers live in the Dark Zone. Most of the animals in this zone are black or red in color due to the lack of light. Occasionally a sperm whale may dive to that level in search of food.

As waves move across the ocean, the water does not move with the waves. Waves lift water particles up and down in a little circle.

5

Ocean waves move in transverse waves. Transverse waves can be observed by gently flicking a jump rope.

7

Large ships ride over most waves. Smaller ships ride up one side of a wave and down the other. A severe storm or hurricane can damage ships by dumping tons of water on them in only a few seconds.

12

10

Ocean water is always moving. The ridges that we see on the surface of the water, called waves, are created by winds.

Volcanoes and earthquakes under the ocean can also create waves.

1

The faster the wind blows and the longer it blows, the greater the fetch. The greater the fetch, the bigger the waves. Strong, steady winds that blow across the ocean eventually produce long, smooth waves called swells.

3

Wonderful Wonders in the Water

Waterspouts are formed when tornadoes or whirlwinds move out to sea. Commonly found in the Gulf of Mexico, most waterspouts have winds up to 600 mph (965 kmh) and water vapor 30 feet (10 m) thick and 400 feet (120 m) high.

16

The manner in which waves break shapes beaches.

14

A bottle floating in the open sea moves up and down on passing waves. It does not move along with the wave. Only the energy of the waves moves forward.

In the deep ocean, storm-driven waves can reach great heights. In 1933 in the North Pacific Ocean, a wave 112 feet (34.2 m) high was measured. That is as tall as a ten-story building.

The size and speed of waves vary depending on the speed of wind, how long it blows, and the fetch. The fetch is the distance a wave travels.

Tsunamis are often mistakenly called tidal waves; however, the moon and Sun's gravitational pull has nothing to do with tsunamis.

The largest waves, called tsunamis, are produced by underwater volcanic eruptions or earthquakes. Tsunamis can reach heights of over 100 ft (15 m) and destroy coastlines.

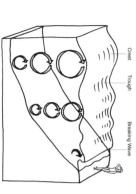

In the same way, the water in a wave moves perpendicular to the wave direction. The water moves up and down, and the wave moves toward the shore.

The top of a wave is called the crest; the bottom of a wave is called the trough. The distance between two crests is called the wavelength. The distance from the crest to the trough is the wave's height.

As waves reach shallow waters near the shoreline, they catch on the ocean floor, causing them to slow down. At the same time, the crest of the wave tries to continue at the same speed until it topples over and becomes a breaker. A breaker is a wave that breaks into foam against a shore or reef.

The lowest water level between the high tides is called low tide. The movement of water which occurs between high tide and low tide allows for a unique environment along the coastline.

A pan of water can represent the ocean. If you tilt it back and forth, you can see that the water at the ends of the pan moves up and down more than the water in the middle.

This is why islands in the middle of the ocean, such as Hawaii, have smaller tides than other coastlines.

If you have ever walked along a shoreline, you have seen the water come high up on the beach.

As early as the 4th century B.C., the Greeks observed a relationship between tides, the moon, and the Sun. However, it was not until Isaac Newton (1642-1727) worked out his theory of gravity that the effects of the moon and Sun on tides were more fully understood.

Newton's theory of gravity was published in Principia Mathematica in 1687.

The side of Earth facing away from the moon is also affected by the moon's gravity, but in a lesser way. This is called a second high tide. Most places along the coastline have two high and two low tides every day.

The time and size of tides can vary, depending on the coastline. In some narrow bays, tides can rise very high. The greatest tidal range on Earth is located in Canada's Bay of Fundy, where high tide may be fifty feet higher than low tide.

Twice a month, the moon and Sun pull at right angles to each other. This produces the smallest tides, called neap tides, because the gravitational pulls of the moon and Sun cancel each other out.

Tides are caused by the gravitational pull of the moon and the Sun.

The daily rise and fall of ocean water is called tides. Tides are caused by the gravitational pull of the moon and the Sun, primarily, and the Sun.

The moon is much smaller than the Sun but is closer to Earth, so its gravitational pull is stronger than that of the Sun. As Earth revolves around the Sun, the waters facing the moon are pulled toward it, causing high tide.

As the tide goes out, water collects in cracks and depressions along the rocky shore.

Twice a month, Earth lines up with the moon and Sun. The gravitational pulls from the moon and Sun team up to produce the biggest tides, called spring tides. Spring tides occur during full moons and new moons.

Note: The term "spring tide" does not relate to the spring season. It comes from a Saxon word which means "to swell."

During another time of the day, the water moves far away from shore. The water moves far away from shore.

If a tide has a wider coastline over which to spread, the rise and fall of tides may be very slight. The tides in parts of the Gulf of Mexico rise and fall only a few inches.

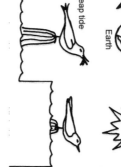

Neap tides occur between spring tides when the moon is only partly visible. This is when the moon is in its first and third quarters.

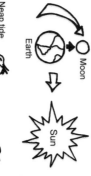

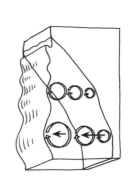

5

There are seven major surface currents and thousands of smaller ones. The smaller surface currents may be cold or warm. Currents move about as quickly as a person can walk.

7

The temperature of the surface currents impacts the climate of land areas nearby. Deep ocean currents are caused by temperature differences and the ocean's saltiness. These currents mix warm and cold waters.

Lots of Science Library Book #11

12

The Peru Current is a cold ocean current that flows northward along the western edge of South America. It is a cold current because it begins near the South Pole, where water is very cold.

Lots of Science Library Book #11

10

Waters in the oceans are constantly moving. Underwater streams, called currents, move through the oceans. Currents flow near the surface of the water and deep in the ocean.

Lots of Science Library Book #11

1

Earth's rotation causes winds to shift.

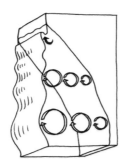

3

Wonderful Wonders in the Water

One type of squid is born with both eyes the same size. As it ages the right eye grows up to four times larger than the left eye.

Lots of Science Library Book #11

16

The ocean's saltiness and temperature causes currents to also flow up and down. Water moves from the depths to the surface and back again.

14

Page 2

Surface currents are caused by a combination of Earth's rotation, winds, heat from the Sun, and land masses. Without Earth's rotation, winds would blow only north and south.

Page 4

A combination of winds and large land masses causes surface currents to flow in a circular motion. In the Northern Hemisphere, currents flow clockwise. In the Southern Hemisphere, currents flow counterclockwise.

Page 6

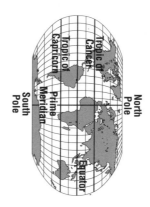

Warm water from the equator is pushed toward the colder polar regions. As a result, the colder water near the North or South Poles is pushed back toward the equator.

Page 8

The Gulf Stream is a warm ocean current that flows along the eastern United States. This is a warm current that begins at the equator, where the water is warm, and flows toward the north.

Page 9

Page 11

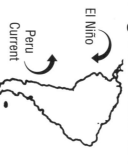

El Niño means "the child" because it usually occurs around Christmas time.

Another current called El Niño, along the west coast of South America, can bring heavy rainfall to some areas and droughts to other areas.

Page 13

Cold water currents flow toward the equator and replace the warmer waters that flow toward the poles. Warm water is less dense than cold water, so warm water is constantly rising above the colder water.

Page 15

Salt makes water more dense, or heavier. As saltier water sinks, it mixes with the surrounding water, making it less salty. The less salty water rises, and the waters continue to mix in this manner.

Page 5

Page 12

Lots of Science Library Book #12

Page 1

The weather is a part of our everyday lives. We plan events and determine the clothes we wear according to the weather.

Lots of Science Library Book #12

Page 16

Wonderful Wonders in the Water

The Portuguese man-of-war has tentacles that can trail for almost 100 feet (30m). This huge animal is made up of almost 100,000 individual animals stuck together.

Lots of Science Library Book #12

Page 7

Earth's tropical oceans hold heat, while the polar waters remain cold all year. Land, however, quickly warms up or cools off depending on the seasons.

Page 10

Heat from the Sun turns water in the oceans, lakes, rivers, and ground into water vapor. Water vapor rises into the atmosphere through evaporation. As water vapor meets cooler air, condensation occurs. The cool air cannot hold the water vapor, and water vapor turns back into water droplets as precipitation.

Lots of Science Library Book #12

Page 3

Lots of Science Library Book #12

Page 14

Storm surges occur when waters rise underneath hurricanes. If a storm surge reaches land during high tide, it can bring devastating coastal floods. Such a surge occurred in Bangladesh in 1970, killing about 500,000 people.

Lots of Science Library Book #12

Page 2

The ocean plays a key role in Earth's atmosphere and weather. The ocean is so vast and holds such a large volume of water that ocean temperatures change slowly.

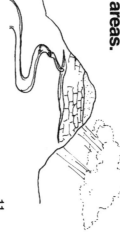

Page 4

Temperatures of land masses may vary from very hot to very cold. Ocean temperatures remain fairly consistent. If you walk along the beach on a sunny day, the sand may feel warm while the ocean water may feel cool.

Page 6

When you return to the beach the same evening, the sand may feel cold. However, the ocean temperature remains fairly stable during the day and night.

Page 8

Page 9

The fairly consistent temperatures of the ocean moderates the temperatures on land.

The effect of the ocean on weather is more obvious in coastal areas, but the ocean also influences all areas of land. Compared to the other planets in our solar system, Earth maintains a fairly consistent temperature, largely due to the ocean.

Page 11

Trade winds, winds blowing almost constantly in one direction, are more regular over oceans than over land. As trade winds move over warm, tropical waters, they pick up moisture and bring heavy rainfall to mountainous areas.

Page 13

Hurricanes are severe tropical storms that form in the southern Atlantic Ocean, Carribean Sea, Gulf of Mexico, and eastern Pacific Ocean. Hurricanes gather heat and energy through contact with warm ocean waters and evaporation from the ocean.

Page 15

Earthquakes under the ocean may cause devastating waves, called tsunamis. Tsunamis move as fast as a jet plane and can reach 100 ft (30 m) high when entering coastal waters. In 2004 in Indonesia, thousands of people were killed by a tsunami.

Animals such as crabs, starfish, sea anemones, mussels, and sea urchins live amongst a variety of seaweeds in crevices or small pools of water. The community of animals, plants, and microorganisms that live in a tidal pool makes up an ecosystem.

Sea anemones have a sticky base that keeps them firmly in place as waves crash against them. They retract their stinging tentacles during low tide to hide themselves and reopen the tentalcles like flowers to ambush prey during high tide.

As tides rise and fall along a rocky coast, an area of land called the intertidal zone is created. During high tide, the intertidal zone is covered with water.

Wonderful Wonders in the Water

Sperm whales can hold their breath for nearly two hours when diving for food. A whale was found with deep sea fish in its stomach that indicate it swam 10,000 feet (3,000 m) beneath the surface of the ocean.

Organisms living in the middle intertidal zone live their lives underwater during high tide and are exposed to air during low tide. Clams bury themselves in the sand and shellfish hide in their shells during low tide.

Tidal pools provide natural habitats for many plants and animals. Some tidal pools have sandy bottoms, allowing creatures such as shrimp and crab to burrow into the sand to avoid predators.

The upper intertidal zone, along with the splash zone, receives the least amount of water exposure. Plants and animals in this zone come into contact with water only during the highest tides and when waves break.

The intertidal zone is commonly divided into four smaller sub-zones: lower zone, middle zone, upper zone, and splash zone. The lower intertidal zone is almost always covered with water. Kelp and other seaweed thrive in these waters but are still vulnerable to strong waves and currents.

Plant life thrives in these waters, providing food and shelter for a wide array of animals. Crabs hide under rocks and use their strong pincers for grasping food. They eat small animals and the remains of dead animals.

Ebb tide, when waters flow back toward the sea, leaves behind small pools of water in the cracks and openings on the rocky coast. These small pools are called tidal pools.

Low Tide

Seaweed have root-like structures that anchor them tightly to rocks. Sea urchins' spines allow them to wedge themselves into cracks and keep predators away.

Sea stars are vulnerable to drying out, so they are found only in this underwater zone and the lower part of the middle intertidal zone. The undersides of their tentacles are equipped with suction-like tubes which help them hold on to the sea floor during the movement of the tides.

Most of the animals living in tidal pools are invertebrates: animals without backbones.

These animals have usually adjusted to living completely in water and they can endure harsh waves. They can also survive in the open air.

Because organisms in these zones must be able to withstand long exposure to the Sun without moisture, the upper intertidal and splash zones are the least populated. The periwinkle, a type of mollusk, secretes mucus to prevent drying out in this harsh environment.

Mussels attach themselves to the sea floor, or to each other, growing into large beds. Sometimes large chunks of mussel beds are torn away by strong waves. Mussel beds provide a natural habitat for snails and worms.

Page 5

Page 12

Penguins are another type of flightless bird that depend on the ocean. Varying in size from the emperor penguin at 3.7 feet (1.1m) to the fairy penguin at 16 inches (41 cm), penguins are known for their streamlined body and dark and white coloring.

Page 7

Page 10

Puffins are sea birds that live in large groups on cliffs high above the seashore. They use their wings to swim underwater and catch small fish in their colorful beaks.

Page 1

The ocean shore is abundant with animals and plants that depend on the daily tides for survival.

Page 3

Page 16

Wonderful Wonders in the Water

Parrot fish secrete a jelly-like substance to build a sleeping bag. This protective bag takes about 30 minutes to create.

Page 14

Shiny feathers overlap to cover a penguin's skin. Penguins have more feathers than most other birds. Each year, usually after the breeding season, penguins will molt, shedding their feathers and growing new ones.

Sea birds are abundant on most coastlines. Sea gulls have a hooked bill, webbed feet, and long wings. They can easily soar over the shallow ocean to find fish for food. These ocean birds also eat shellfish and insects. Many types of sea birds search the ocean shore for various types of food.

With wing spans of 6 to 8 feet (1.8 – 2.4 m), pelicans are huge birds that line the shores of various oceans around the world. These unique birds use a plunge-diving technique to catch fish and have a pouch-like mouth to trap the fish.

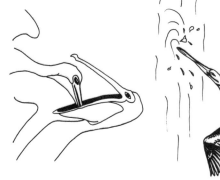

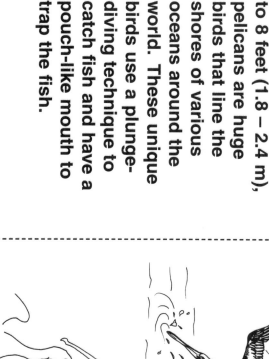

Seaweed is a type of algae. It has root-like organs, called holdfasts, which cling to rocks. Seaweed is tough and rubbery allowing it to survive time on shore and return to the ocean during high tide.

Hermit crabs are soft bodied animals that use the discarded shells of snails as their home. They have a huge right claw that is used to cover the entrance of the shell when predators attack. When the hermit crab outgrows its shell, it must find a larger discarded one on the shore or ocean floor.

Bar-built estuaries are formed when sandbars build up along the coastline partially cutting off the ocean waters. Bar-built estuaries are usually shallow, with reduced tides. This type of estuary is common along the Texas and Florida Gulf coasts.

Earth's crust is constantly in motion, causing large cracks and folds in the crust. At times land sinks, or subsides. Tectonic estuaries are formed when the sea fills in the hole or basin formed by the sinking land. San Francisco Bay in California is an example of this type of estuary.

Estuaries are partly enclosed bodies of water that contain salty ocean water and sometimes freshwater from a nearby river. The colors of estuaries are green and brown, as most are shallow bays with little wave movement.

Fjords are formed when valleys once cut by glaciers are flooded by the ocean waters. They are usually narrow with steep sides. Fjords are found in areas that have been covered by glaciers such as Chile, New Zealand, Canada, Alaska, and Greenland.

Wonderful Wonders in the Water

In Norway, the Geiranger Fjord lies between steep mountain slopes and is flooded by the Atlantic Ocean. This fjord includes many waterfalls, including "The Seven Sisters."

Many marine animals begin their lives in the safe, calm shelter of an estuary. While most of the marine life in an estuary are visitors, small fish, mud snails, and oysters may spend their entire lives in the estuary.

Along the east coast of North America, one of the strangest estuarine residents is the 'living fossil' known as a horseshoe crab. With four eyes, five pairs of legs, a spear-like tail, and blue blood, this marine animal is not a true crab but a close relative of scorpions and spiders. There are only five living species of horseshoe crab today.

Estuaries are divided into four types, depending on the manner in which they are formed:

1. Coastal Plain Estuaries
2. Tectonic Estuaries
3. Bar-built Estuaries
4. Fjords

Coastal Plain Estuaries are formed when the ocean level rises and fills an existing river valley. The valleys are usually shallow with gentle sloping bottoms. This type of estuary is common throughout the world. Chesapeake Bay in Maryland and the harbor in Charleston, South Carolina are examples of this type of estuary.

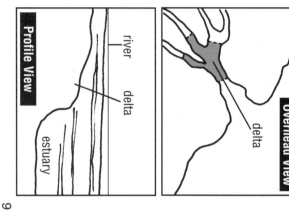

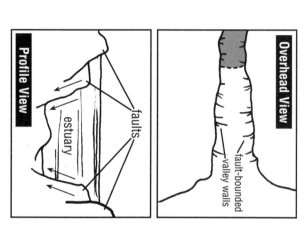

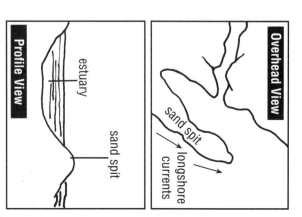

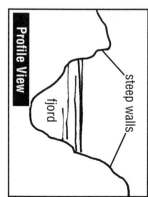

Lobsters are called "the fierce cannibal" because they will attack another lobster, especially if its body is soft, while a new case is hardening.

5

Yet, even when a crustacean has its hard outer case, many predators of the sea can easily crack the case and eat the animal.

7

Other types of mollusks have no shells, such as the octopus, squid, and cuttlefish.

12

Bivalves are two-shell mollusks. The shells are joined by a hinge. As an oyster grows, its shell must also grow. The mantle is an organ that produces the oyster's shell, using minerals from the oyster's food.

10

Crustaceans are probably best known for their hard outer cases. This group of animals includes lobsters, crab, and shrimp. Like fish, crustaceans breathe through gills.

1

3

Wonderful Wonders in the Water

Small cleaner wrasses clean off dead skin and parasites from the bodies of larger fish. Even moray eels stay very still while the wrasses clean leftover food from their sharp teeth.

16

Sponges differ from all other invertebrates because they have no true tissues or organs. Sponges are a network of openings and canals that connect to open pores on their surface. Sponges feed by pulling in water through their pores and filtering out the nutrients

14

Lobsters are born from eggs. When the larva, or baby lobster, hatches from the egg, it floats freely with the plankton. Soon the larva moves to the bottom, where it develops its first casing.

As a crustacean grows, its hard case must be removed and discarded. A new one takes time to harden. During this period, the animal is without its primary means of protection and is open to attack from predators.

Although vulnerable during the hardening of their new cases, many crustaceans have other means of protection they depend upon. The claws of many crabs are capable of exerting hundreds of pounds of pressure. The mantis shrimp can break the glass of an aquarium or split a man's thumb to the bone with one strike.

Mollusks are a group of soft-bodied animals that includes snails, clams, and sea slugs. A hard external shell is the most common characteristic of a mollusk.

Mollusks are found in saltwater and freshwater environments.

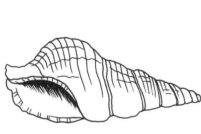

Snails are univalves, which means they have one shell.

A natural pearl begins when sand or another unwanted substance enters the oyster between the mantle and the shell. The mantle's reaction to this irratation is to cover the substance and protect itself. The foreign substance is covered by many layers and eventually forms a pearl.

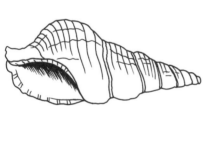

Although ocean sponges may look like plants, they are brilliantly colored sea animals. Most are small, but some can grow over 6 feet in diameter.

Page 1

There are three types of fish; bony fish, cartilaginous fish, and jawless fish.

Page 5

Cartilaginous fish are covered by a rough surface skin instead of scales. Their fin pattern is similar to that of a bony fish, although their tail fin is different.

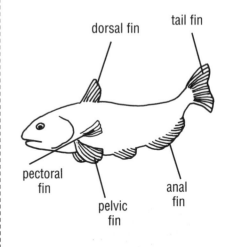

Page 3

Cartilaginous fish have five gill slots. Because they do not have a swim bladder, they must swim continuously or drown.

Page 10

Cartilaginous fish have a skeleton made of cartilage; the flexible substance found in the ears and nose of humans. Their bodies are usually wide and flat. Cartilaginous fish include sharks and rays. See *Lots of Science Library Book #20* for more information.

Page 7

The blood vessels absorb oxygen from the water and then the water exits behind the gill covering.

Like all animals, fish need oxygen to breathe. Most fish receive oxygen from the water through gills. Gills hold many feather-like structures, called filaments, which contain blood vessels. Oxygen-rich water enters the fish's mouth and is forced over the gills.

Bony fish make up about 96% of all fish species. Bony fish have a skeleton of bone, with a backbone and ribs. Most bony fish are covered with scales and one pair of gills covered by protective plates.

Jawless fish are the simplest class of fish and all are extinct except hagfish and lampreys. The bodies of hagfish and lampreys are long, tube-like, slimy, and lack scales and paired fins.

Bony fish have a swim bladder, or air bladder, located in the dorsal portion of their body. The swim bladder, which is filled with gases, adjusts the density of the fish so it is equal to the surrounding water. This allows the fish to be still and not drown.

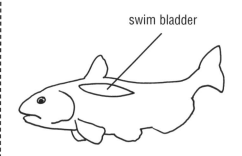

The paired fins consist of pectoral and pelvic fins. Their tail fin is made up of upper and lower fins. On the top of their body is the dorsal fin and on the bottom is the anal fin.

Wonderful Wonders in the Water

A starfish can escape from danger by leaving some of its arms behind. All starfish can grow new arms and some can grow a new body from a tiny piece of arm.

Sea anemones are characterized by a symmetrical body, usually with stinging tentacles, and a central mouth. They usually attach to rocks, coral, or shells.

The vivid colors on these fish make them beautiful to observe. Many people scuba dive for the purpose of watching these amazing fish. Others keep saltwater aquariums so these fish can be enjoyed daily.

Tubeworms form a hard-shelled tube for protection. Tubeworms that live near hydrothermal vents live longer and grow faster than ones that live in the cool areas of the ocean floor.

Stripes allow a fish to camouflage itself against coral and confuse a predator.

Sea anemones are relatives of corals and sea fans. Jellyfish are also members of this sea animal group.

All echinoderms move with thousands of tiny tube feet. Many have suction cups on the ends. Many of the urchins have extremely sharp spines used as a means of protection.

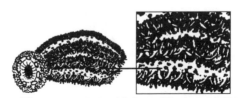

Flatworms have flattened bodies and look like chewing gum as they search for food on the rocks.

Feather duster worms have feathery tentacles on top that are used to filter nutrients from the water. When threatened by predators, they quickly withdraw into their tube homes.

Marine fishes are the most colorful and varied group of sea animals.

The tentacles can sting, paralyze, and entangle small marine animals.

Marine fishes are an important part of coral reefs throughout the oceans. See *Lots of Science Library Book #19* for more information on this ecosystem.

Echinoderms are a group of animals that includes starfish, urchins, feather stars, and sea cucumbers. They are characterized by symmetry and a central mouth.

Several species of starfish push their stomachs into their prey to digest them.

Marine worms are a large group of sea animals. Sea worms are segmented worms that look much like a common earthworm.

Reddish colors can look black underwater, helping a fish go unseen. These fish are unique and colorful for their protection.

Wonderful Wonders in the Water

At night the surface of the Indian Ocean sparkles with light made by tiny sea plants called din flagellates. The lights are bright enough to read by.

16   Lots of Science Library Book #19

Beneath the clear, warm waters of tropical oceans, the beautiful world of the coral reefs is found. With shapes resembling mushrooms, antlers, fluted pillars, and brains in brilliant colors of yellows, purples, and oranges, coral reefs are the favorite places for snorkelers and divers to visit.

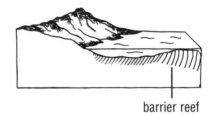

Lots of Science Library Book #19   1

Barrier reefs are separated from the coastline by a lagoon. Barrier reefs grow parallel to the coastline and are usually wide and long.

barrier reef

12   Lots of Science Library Book #19

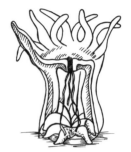

5

The combination of tropical currents, temperature and waterborne nutrients make this an ideal place for reef building. The Great Barrier Reef is made up of thousands of small reefs connected or built near one another. The structure is so large it can be viewed from space.

14   Lots of Science Library Book #19

3

Atolls form in a ring shape sometimes around or near an island. The water inside the reef is called a lagoon. Atolls can rise from the deep sea or can be found on the continental shelf.

10   Lots of Science Library Book #19

A coral reef consists of thousands of polyps. In addition, the coralline algae cement the various corals together with compounds of calcium. Other organisms such as tube worms and mollusks contribute their hard skeletons to the reefs, founding unique reefs throughout the tropical oceans.

7

Polyp skeleton is made of limestone. The polyp's cup-shaped body is ringed with stinging tentacles so it can catch food from the water.

atoll

lagoon

Anemones, sponges, mollusks, urchins, sea snakes and colorful marine fish find food and shelter in the coral reefs. In fact, the corals themselves are marine animals.

Coral reefs found in the Pacific Ocean are far more varied and colorful than the ones located in the Atlantic and Caribbean Oceans.

In addition to providing a beautiful home for marine life, coral reefs also protect the coastlines from erosion. There are three types of reefs:
 1. fringing reefs
 2. atolls
 3. barrier reefs.

Fringing reefs grow in shallow waters and border the coast very closely or are separated from the coast by a narrow stretch of water.

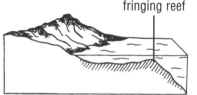

fringing reef

Although the coral reef may look like a large mass of rock and growing plants, it is a living colony of tiny sea creatures called coral polyps. Coral polyps grow by dividing, and produce a hard outer skeleton to protect their soft bodies.

The Great Barrier Reef, off the coast of Australia, is the world's largest coral reef system. It stretches about 1,250 miles (2012 km) along the coast and ranges from 10 – 60 miles (16 – 96 km) from the shoreline.

Wonderful Wonders in the Water

The most powerful electric sea fish is the black torpedo ray. Using electric shocks to kill prey and as defense, this ray can generate enough electricity to run a television set.

Although sharks are viewed as vicious man-eaters, only a small handful are a threat to man.

Sharks have from 1 to 100 babies at a time. Sharks with pups that grow inside the mother have fewer babies at a time than ones that lay eggs outside the body. Sharks do not take care of the babies after they are born; however, they search for a safe place where they can lay eggs or give birth.

Sharks are part of a family of fishes known as cartilage fishes. Rays are also a member of this group. In fact, rays are actually flattened out sharks.

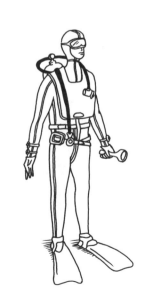

The basking shark, the whale shark, and the megamouth shark are quite harmless. These huge sharks eat zooplankton. To do this, they swim forward with their mouths wide open. Gill rakers at the back of the throat strain the tiny food from the water.

Shark teeth grow in parallel rows. Rows of replacement teeth grow behind the outer teeth. A nurse shark, on average, replaces each front row tooth every ten days to two weeks.

A shark may grow and use over 20,000 teeth in its lifetime.

Sharks have the most powerful jaws on Earth. The upper and lower jaws of a shark move when it bites down. A shark bites with its lower jaw first and then its upper jaw. It tosses its head back and forth to tear loose a piece of meat, and then swallows it whole.

Baby sharks are called pups. There are three ways that sharks are born: eggs are laid (like birds), eggs hatch inside the mother and then are born, or pups grow inside the mother (like mammals).

Most shark attacks are probably a case of mistaken identity. A diver in a wet suit may look like a sea lion, a favorite food for larger sharks.

Rays move gracefully through the water as easily as birds fly through the air. Some rays are capable of inflicting painful stings with their tails.

A shark's skeleton is made mostly of cartilage which allows great flexibility in movement. While swimming, many sharks can switch directions very quickly.

A shark's skin is made of denticles instead of fish scales. Denticles are made like hard, sharp teeth and help protect the shark from injury.

Each type of shark has different shaped teeth depending on its diet. This is the mouth of a Great White Shark. The sharp, pointy teeth indicate that it is a carnivore living off a diet of fish, dolphins, seals, turtles, and sea gulls.

Sharks are very successful predators because they have keen senses. Two-thirds of a shark's brain is dedicated to its strong sense of smell. Some sharks have eyes that allow them to see clearly in the murky water.

Page 1

One of the most well known marine reptiles is the sea turtle.

Page 5

Sea turtles live and mate at sea. Males rarely return to land after they enter the ocean shortly after birth. Females migrate long distances to reach their breeding sites. Sea turtles lay their eggs on land. Most females return to the same beach each time they are ready to nest.

Page 12

Sea snakes are venomous, but not aggressive except during mating season.

Page 14

Sea snakes feed mainly on fishes. Some of them are active hunters, hiding in crevices and holes awaiting prey.

Page 3

Most turtles feed on jellyfish, sponges, soft corals, crabs, squids, and fishes. Some turtles also feed on sea grasses and algae.

Page 10

Many turtles drown accidentally because they get tangled in nets. Sometimes, bits of plastic are mistaken for food. Once it is swallowed, the plastic blocks the turtle's breathing and digestive processes.

Page 7

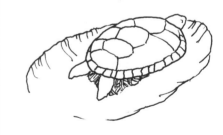

They can be seen on the beaches late at night digging a deep hole in the sand. The female sea turtle crawls to a dry part of the beach and flips loose sand with her flippers. By rotating her body and digging with her hind flippers a suitable nest is ready for her eggs. Between 50 and 150 eggs are deposited and covered over with sand.

About 50 species of sea snakes live in the ocean. Sea snakes have flattened tails for swimming, and valves over their nostrils which close underwater.

There are only seven different species of sea turtles throughout the world. They range in size from 2 feet to over 6 feet in length. Sea turtles are known to feed and rest throughout a typical day.

Like all reptiles, sea snakes shed their skins; but sea snakes shed much more frequently than land snakes. Sea snakes shed their skins as often as every two weeks.

About 2 months later, tiny turtles dig their way to the surface and scramble to the sea. Only 1 in a 1,000 survive the predators and grow to adulthood.

Turtles are one of the world's most endangered species. In some areas, people eat turtle meat and eggs. Turtle shells are used for jewelry and turtle oils are used in cosmetics.

Sea turtles breathe air, but they can spend as much as 3 hours underwater. Most turtles can dive down to depths of over 1,640 yards (1,500 meters). These turtles can sleep on the surface of the water or on the bottom wedged under rocks.

Sea snakes are different from eels. Sea snakes do not have gill slits. They must breathe air and therefore are found in shallow water. The sea snake has a blunt flattened head and scales.

Sea otters hunt for food on the ocean floor, always returning to the surface to eat while floating on their backs. Sea otters live in the Pacific Ocean and rarely leave their water home.

Like all mammals, marine mammals are warm-blooded, give birth to living young, and breathe air through lungs. Marine mammals include dolphins, whales, and seals.

At times, orcas hunt like a pack of wolves. A pod of orcas can surround a large whale and overcome it with no difficulty. Sometimes a rogue pod will attack and kill other orcas.

Bottlenose dolphins hunt near the surface of the water, eating mostly fish and squid. They have many pairs of sharp, pointed teeth distributed in both the upper and lower jaws. Bottlenose dolphins are very social animals and live in pods of up to 12.

The blue whale is Earth's largest animal, reaching up to 100 feet (30 m) and weighing up to 300,000 pounds (136,000 kg). As a free swimmer, it travels through the Sunlite and Twilight Zone. Its diet includes krill, a type of zooplankton found near the water's surface.

Sperm whales use sonar in dark depths to locate prey. Sound waves move toward an object; as the sound waves echo back, sperm whales can determine the object's position, distance, and size.

Experienced observers can recognize the blue whale by the size and shape of its cloud of air.

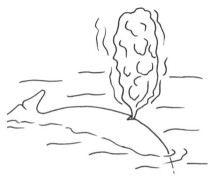

All marine mammals must come to the surface of the water to breathe. The blue whale rises to the surface, emitting a spray of water from a blowhole located at the top of its head. The warm air of the whale's body contacting the cool air outside forms a cloud of water vapor.

Orcas, also known as killer whales, live in family groups called pods. Their diet includes whales, dolphins, seals, sea lions, penguins, turtles and fishes.

Marine mammals show remarkable abilities to communicate and learn. Their natural lives are spent in close family groups caring for their young and each other. Their songs can be heard echoing for miles beneath the waves.

Dolphins can dive down to more than 1,000 feet (300 m) and can jump up to 20 feet (6 m) out of the water. A bow rider is a dolphin that hitches a ride in the bow wave of a ship. The dolphin surfs using the pressure from the front of a moving ship.

The humpback whale, another baleen whale, sometimes hunts by blowing bubbles as it circles near the surface. The ring of bubbles traps the shrimp-like krill from escaping. When the whale surfaces, it swims through the mass of krill with its mouth wide open.

Sperm whales are the largest of the toothed whales. Sperm whales have large heads, about one third of their body length, with an almost hidden lower jaw. Their diet includes squid, bottom fish and sharks.

The blue whale has a huge mouth that sucks in large volumes of food-laden water. Baleen plates hang from the upper palate to strain food from the water. A blue whale has 600 - 800 plates that range from 20" - 40" long.

The bottlenose dolphin is the most familiar of the dolphin family. It can grow to be 12 feet (3.7 m) long.

Wonderful Wonders in the Water

Northern elephant seals are the only mammals that migrate twice a year. The average elephant seal travels over 21,000 miles (33,800 km) each year in these migrations.

16   Lots of Science Library Book #23

Another group of marine mammals are the pinnipeds. The word *pinniped* means fin-footed. This group of mammals includes seals, sea lions, and walruses.

Lots of Science Library Book #23   1

California sea lions feed on squid, fish, and small sharks.

12   Lots of Science Library Book #23

Most seals cannot use their rear flippers to walk on land or ice so they appear awkward when moving on these surfaces. However, seals can 'toboggan' rapidly across the ice. Most of the world's seals live close to Antarctica and the Arctic Circle.

5

Walruses live most of their lives in the icy ocean water of the Arctic. During the spring, walruses leave the water to breed. For a few months, walruses are found on the icy shores or floating ice near the northern continents.

14   Lots of Science Library Book #23

About 18 species of seals inhabit the ocean. They are found in all parts of the ocean and a few freshwater lakes. Seals do not have an ear that extends from the head, but they can hear well.

3

Another distinct population of California sea lions is found at the Galapagos Islands. A third population once lived in the Sea of Japan but became extinct, probably during World War II.

10   Lots of Science Library Book #23

The largest seal, the elephant seal, can weigh more than 5,000 pounds (2,270 kg). The elephant seal has a head with a nose resembling a short elephant trunk.

7

Harbor seals can sleep with their bodies submerged in the water and only the tip of their noses out to breathe. This position is called bottling. Harbor seals have a diet of fish, squid, and crustaceans.

6   Lots of Science Library Book #23

California seal lions are social animals living in large groups. They pack together very closely, even floating together in "rafts." Sometimes these animals will jump out of the water or "surf" the breaking waves.

11

Many of these mammals live in the icy waters of the ocean. This is possible because of their thick skin and the thick layer of oily blubber located under the skin.

2   Lots of Science Library Book #23

The walrus feeds along the ocean floor, using its tusks to rake the seabed for food. Its favorite food is mollusks, especially clams. With 18 teeth walruses can eat 3,000 – 6,000 clams in a feeding, usually sucking the clam out without breaking the shell.

15

Elephant seals crawl on land with rhythmic belly flops because their hind flippers are turned underneath their bodies. In the water, the elephant seal is considered the second best diver in the ocean, reaching depths of 5,000 ft (1,500 m).

8   Lots of Science Library Book #23

California sea lions are known for their playfulness, intelligence, and noisy barking. California sea lions live from Vancouver Island, British Columbia to the southern tip of Baja California in Mexico.

9

Seals have a torpedo like shape and propel themselves through water using their hindflippers. The front flippers are short with sharp claws and are used to steer the seal through the water.

4   Lots of Science Library Book #23

Walruses are known for their large size, big tusks, and many whiskers.

13

Wonderful Wonders in the Water

About 10% of the world's coral reefs are lost due to damage from humans. *Reef Relief* in Key West, Florida is working to keep our coral reefs safe from pollution, boat anchors, and tourists who do not respect the value of our coral reefs.

16

Many products we use every day are taken from the ocean. Two important ones are oil and natural gas. Currently, almost one-fifth of the world's oil and natural gas comes from beneath the ocean floor.

oil and natural gas

1

The ocean provides many other important products we use every day including salt and tin that is mined near Southeastern Asian countries. Manganese is found on the ocean floor as are other useful minerals.

12

An oil platform may stay in place for 25 years or longer.

5

Despite our great love of the ocean, it is currently used as a dumpsite for many cities and individuals.

14

3

If these plastic nets are lost, they cannot rot so they float throughout the ocean trapping and killing marine life for years.

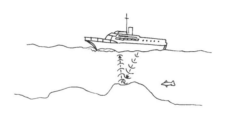

10

Every year nearly three million tons of oil enter the ocean through oil tanker spillage, production wells, and runoff from the land. Although the ocean can recycle many pollutants, it can take many years to do so. The immediate result is the loss of marine wildlife.

7

While the oil taken from the ocean is important, the pollution caused by oil spills results in damage to marine habitat, vegetation, and shoreline wildlife.

Overfishing has forced Florida fishermen to go further out to sea to locate deep sea fish. The Grand Banks, the pride of New England fishing for centuries, are closed due to overfishing.

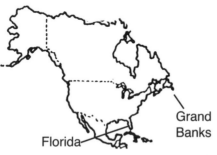

Oil companies send modern day explorers to search for oil beneath the ocean floor. Holes are drilled deep into the soil under the ocean. When a large amount of oil is discovered, an oil platform is placed over the area.

Not only does the garbage ruin the environment for marine life, but the results of such pollution can trap and kill animals in many other ways.

A vast array of delicious seafood is harvested from the ocean. Many people throughout the world depend on fishing for their livelihood. In fact, fish provide the most protein for humans to eat.

Radar and sonar allow large shoals of deep sea fish to be located easily, causing a shortage of fish stock throughout the world. Modern plastic nets are lightweight so they can be largeenough to catch many more fish than in the past.

The oil platform is constructed on land and towed to sea. Once in place, it is anchored to the ocean floor with steel supports. The drilling equipment on the platform is designed to extract oil from the ocean floor. The platform also includes living quarters so the workers can work for several weeks at a time.

The ocean is a playground for millions each year. Whether it is a seashore visit or a cruise with diving and fishing, people have always valued trips to the ocean.

Note: The owner of the book has permission to photocopy the *Lab Log* and *Graphics Pages* for his/her classroom use only.

Graphics Pages

The Great Science Adventure Series

Investigative Loop™

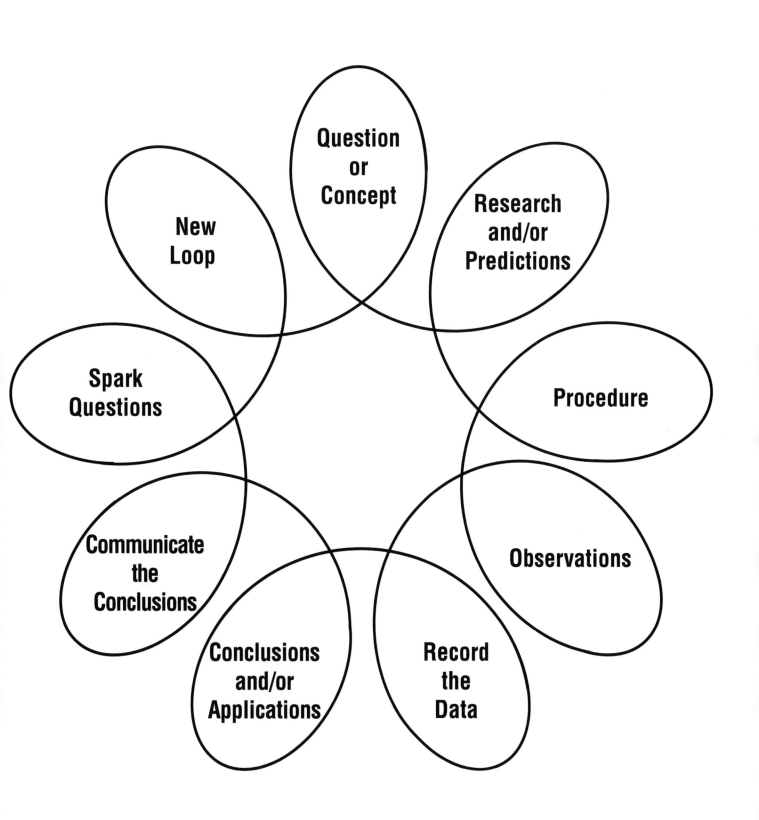

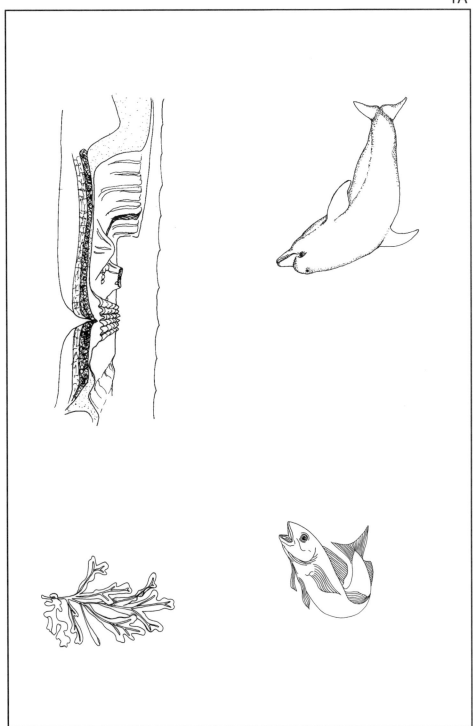

1B

glue

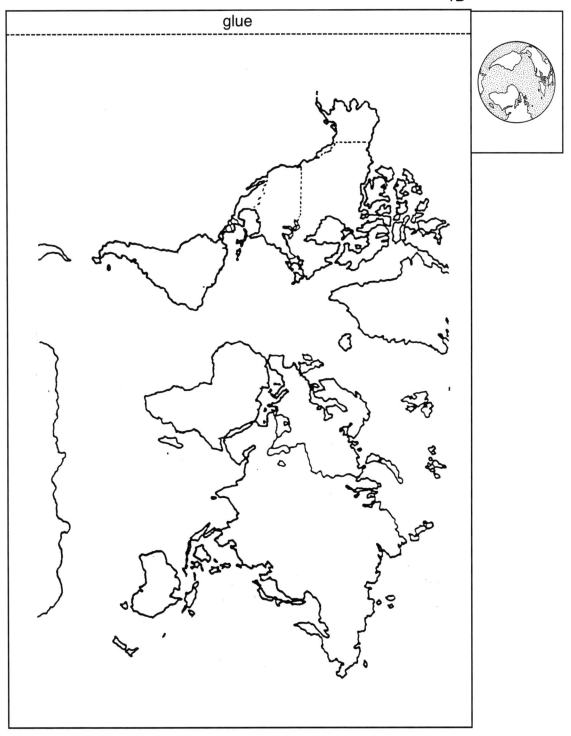

Lab Graphic 1-1

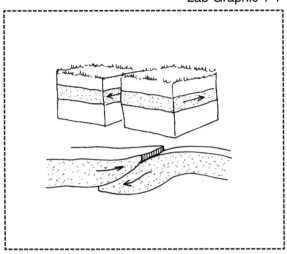

Lab Graphic 2-1

Lab Graphic 5-1

Lab Graphic 5-2

Lab Graphic 7-1

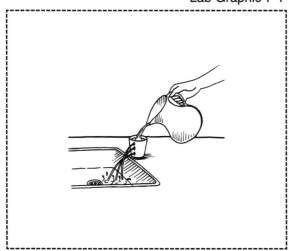

Lab Graphic 9-1

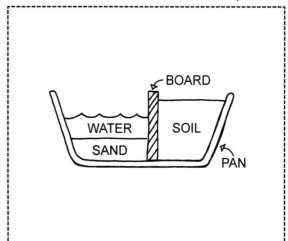

Lab Graphic 11-1

Lab Graphic 12-1

2A

glue

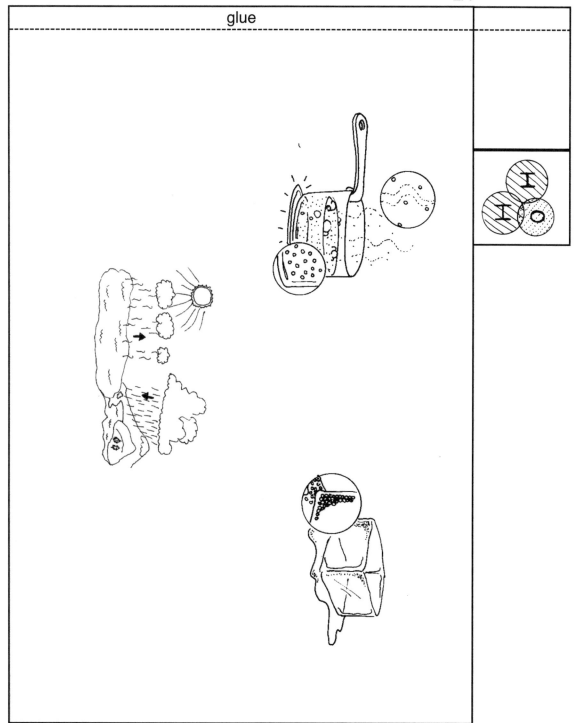

4A

glue

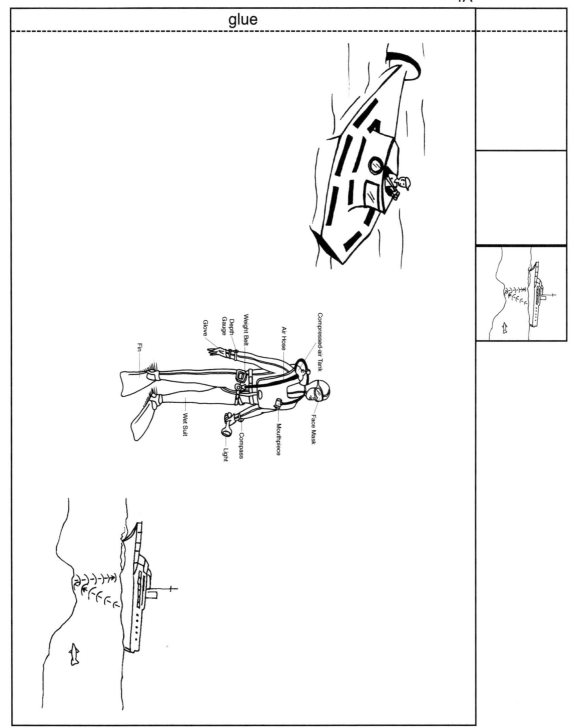

5A

glue

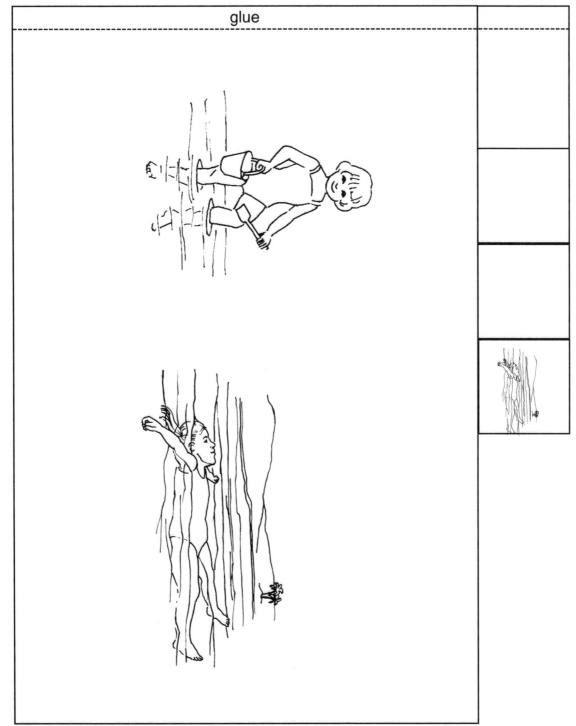

6A

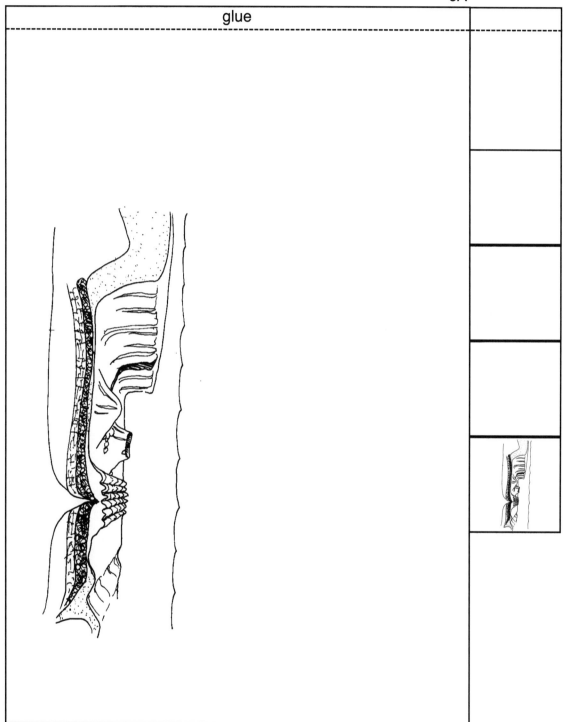

glue

7A

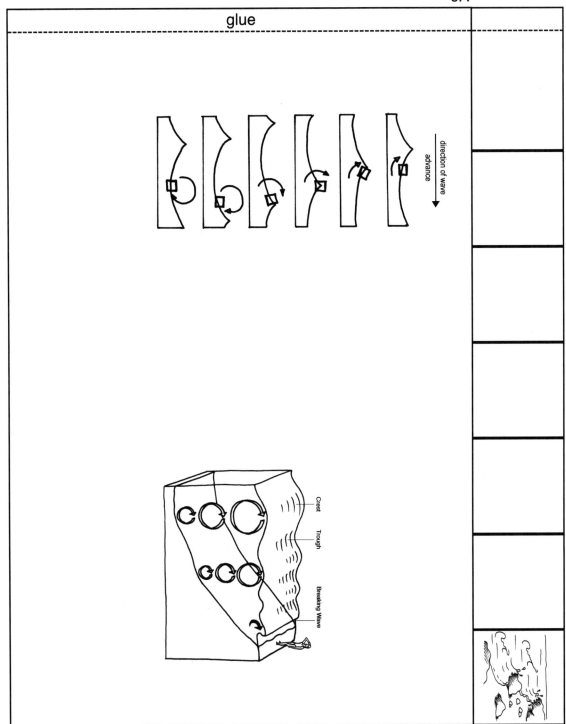

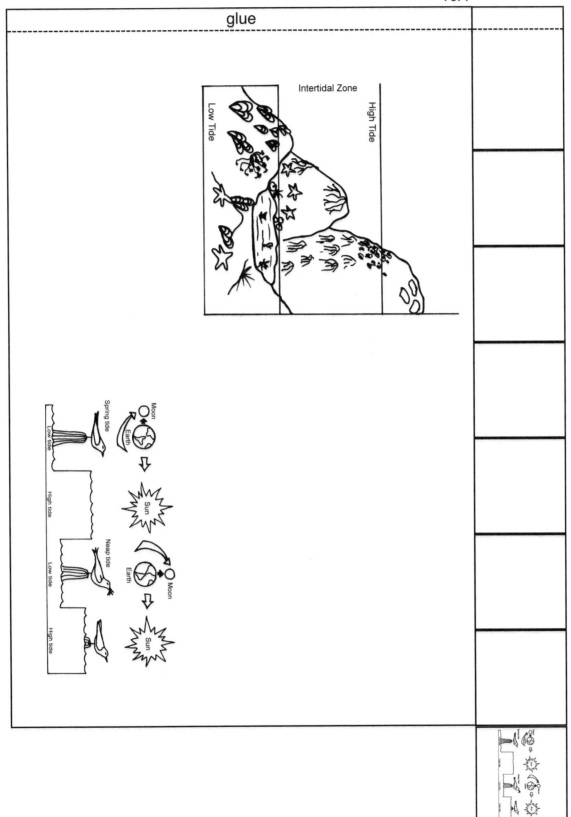

11A

glue

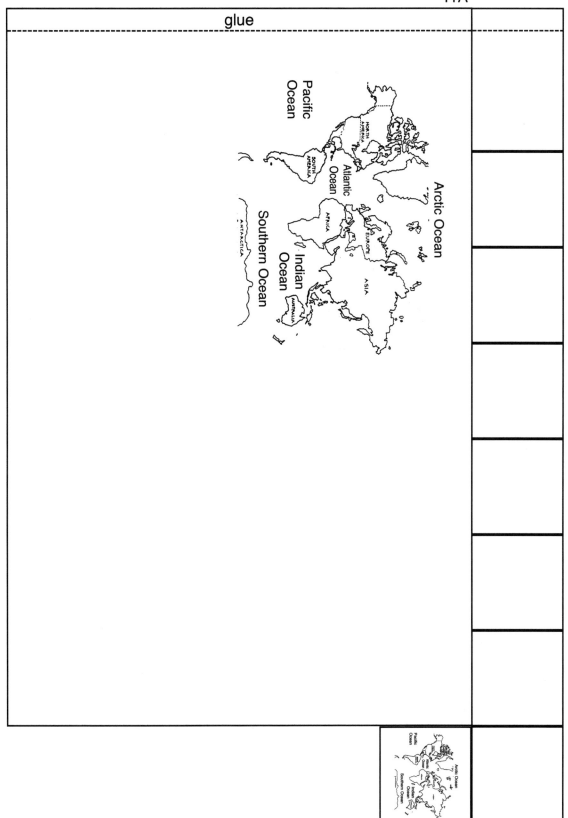

glue

12A

13A

glue

glue

14A

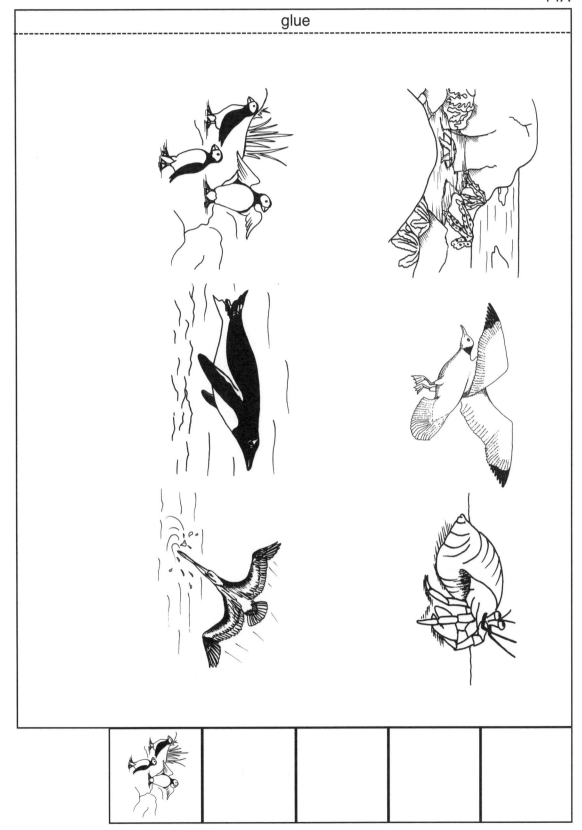

14B

14C

16B

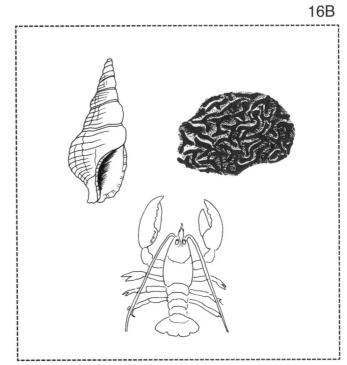

17B

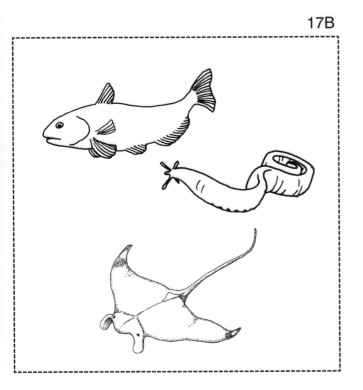

18B

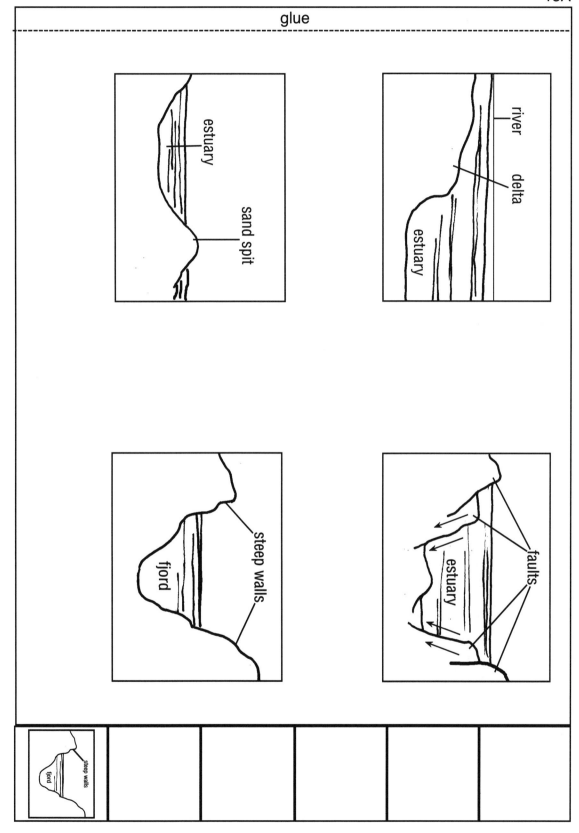

16A

glue

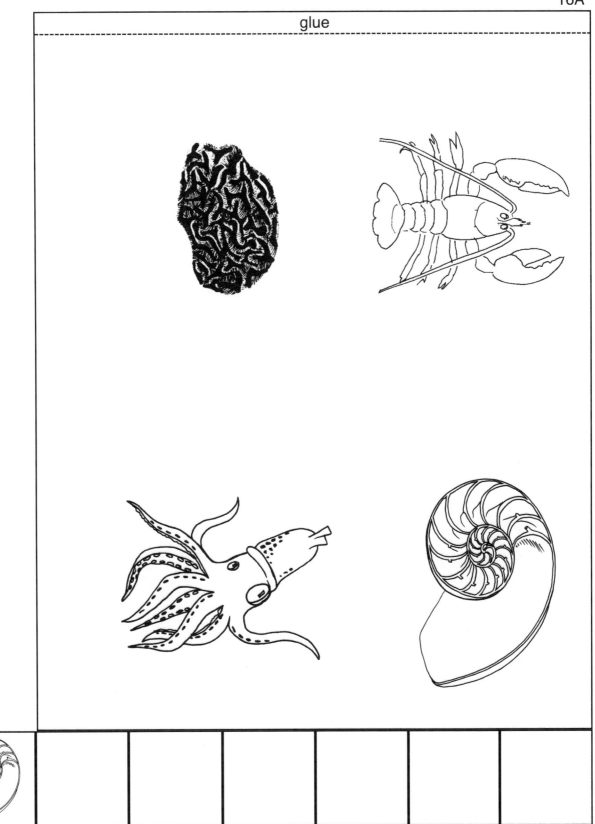

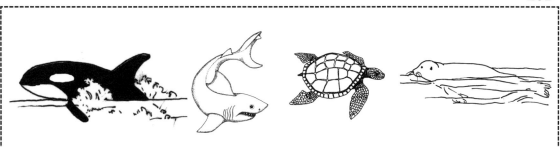

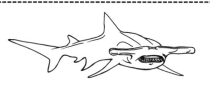 **Sharks and Rays**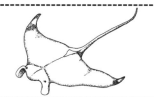

 **Marine Reptiles**

 **Marine Mammals**

 **Pinnipeds**

17A

glue

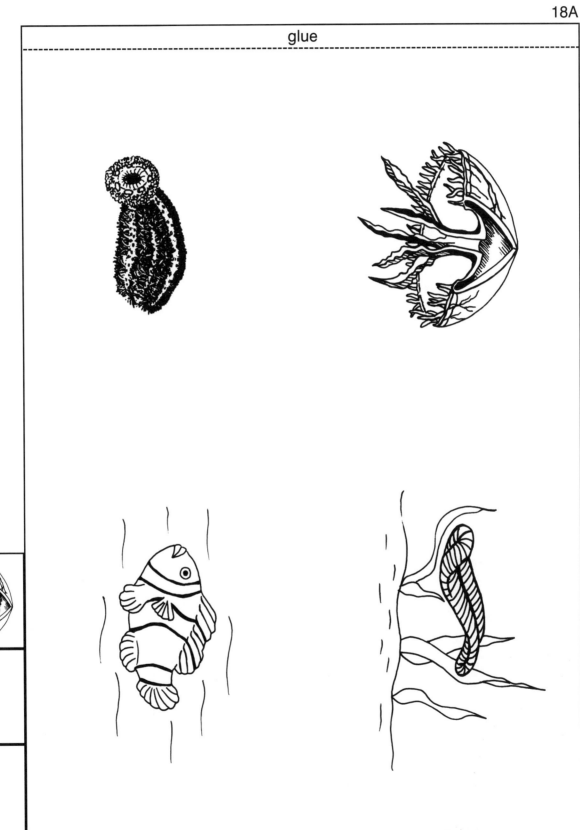

glue

19A

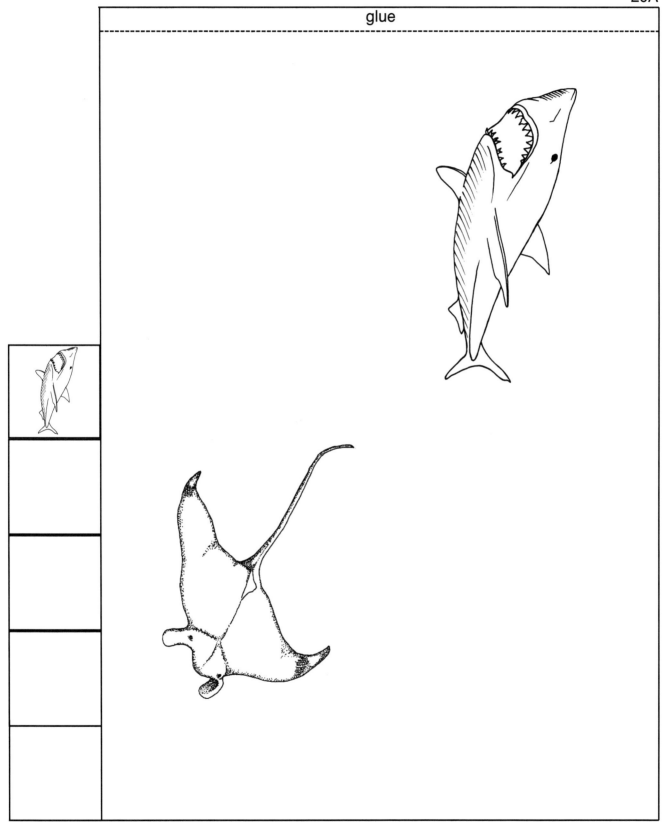

22A

glue

23A

glue

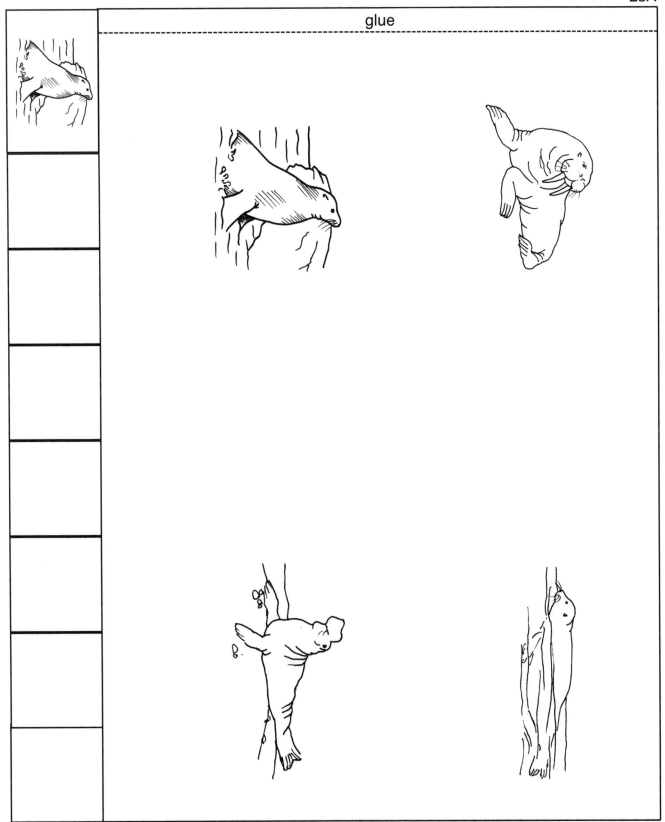

24A

24B